Second Edition

Geochemical
Analysis

Second Edition

Geochemical
Analysis

Rabindra Nath Hota PhD, DSc
Professor
PG Department of Geology
Utkal University, Vani Vihar
Bhubaneswar, Odisha

CBS

CBS Publishers & Distributors Pvt Ltd

New Delhi • Bengaluru • Chennai • Kochi • Kolkata • Mumbai
Hyderabad • Jharkhand • Nagpur • Patna • Uttarakhand

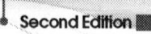

Second Edition

Geochemical Analysis

ISBN: 978-93-89565-90-4

Copyright © Author and Publisher

Second Edition: 2021

First Edition: 2011

Reprint: 2012, 2019

Published by Satish Kumar Jain and produced by Varun Jain for

CBS Publishers & Distributors Pvt Ltd

4819/XI Prahlad Street, 24 Ansari Road, Daryaganj, New Delhi 110 002, India
Ph: 011-23289259, 23266861, 23266867   Website: www.cbspd.com
Fax: 011-23243014   e-mail: delhi@cbspd.com; cbspubs@airtelmail.in
Corporate Office: 204 FIE, Industrial Area, Patparganj, Delhi 110 092

Ph: 011-4934 4934   Fax: 011-4934 4935   e-mail: publishing@cbspd.com; publicity@cbspd.com

Branches

- **Bengaluru:** Seema House 2975, 17th Cross, K.R. Road, Banasankari 2nd Stage, Bengaluru 560 070, Karnataka
  Ph: +91-80-26771678/79   Fax: +91-80-26771680   e-mail: bangalore@cbspd.com
- **Chennai:** 7, Subbaraya Street, Shenoy Nagar, Chennai 600 030, Tamil Nadu
  Ph: +91-44-26680620, 26681266   Fax: +91-44-42032115   e-mail: chennai@cbspd.com
- **Kochi:** 42/1325, 1326 Power House Road, Opp. KSEB, Ernakulum, Kochi 682018, Kerala
  Ph: +91-484-4059061-65   Fax: +91-484-4059065   e-mail: kochi@cbspd.com
- **Kolkata:** 6/B, Ground Floor, Rameswar Shaw Road, Kolkata 700 014, West Bengal
  Ph: +91-33-22891126, 22891127, 22891128   e-mail: kolkata@cbspd.com
- **Mumbai:** 83-C, Dr E Moses Road, Worli, Mumbai 400018, Maharashtra
  Ph: +91-22-24902340/41   Fax: +91-22-24902342   e-mail: mumbai@cbspd.com

Representatives

- **Hyderabad** 0-9885175004   • **Jharkhand** 0-9811541605   • **Nagpur** 0-9421945513   • **Patna** 0-9334159340
- **Pune** 0-9623451994   • **Uttarakhand** 0-9716462459

Printed at Sanjay Printers, Pvt. Ltd., Patparganj Industrial Area, Delhi, India

Foreword

The lack of a monograph on geochemical analysis dealing comprehensively on the subject was felt by the students, researchers and teachers. The subject of geochemical analysis was being dealt separately, i.e. procedures for analysis was dealt separately for water, rocks, coal, etc. in different books/monographs. Hence, the students and researchers in geochemistry had to face the arduous task of searching a number of books to help themselves.

The valiant effort of Dr RN Hota to write a book *Geochemical Analysis* encompassing all the geological materials like rocks, minerals, water, coal, etc. is highly commendable. He not only has enumerated the detailed procedure for analysis, but also has dealt with the instrumentation details. Moreover, in some cases more than one method has been suggested for a single parameter by him leaving it to the convenience of the worker. The book will be of help to both the students as well as the researchers because of its lucidity and comprehensive nature.

I recommend the book to the students, researchers and consultant agencies who are interested for geochemical analysis of geological samples including that of water.

HK Sahoo
Professor
PG Department of Geology
Utkal University, Vani Vihar
Bhubaneswar-751004
Orissa

Preface to the Second Edition

The book *Geochemical Analysis* published in January 2011 got reprinted in 2012, which indicates the overwhelming response from the readers, particularly the students for whom it has been written. I thank all the readers for their support. In this edition, the typographical errors of the first edition have been eliminated. In addition, many of the paragraphs have been rewritten and rearranged for better readability. Particularly the chapter 'Analysis of water samples' has been revised by addition of suitability of water in agriculture and industries. Formulae of SAR, RSC, PI, %Na, PS, KR, MR and corrosive ratio and their interpretations have also been included. Graphical representation of geochemical data, standard diagrams and preparation of reagents are some significant additions. Two chapters viz., 'Analysis of soil sample', and 'Analysis of air sample', have been added to make the book complete in many aspects.

I sincerely appeal to the readers of this book to suggest for betterment in future. They are requested to send their feedbacks and suggestions to my e-mail address (rnhota@yahoo.com).

Rabindra Nath Hota

Preface to the First Edition

The thought of writing a textbook in geology crept into my mind in 1987, when I started my teaching career as a lecturer in the undergraduate section of the Department of Geology, Utkal University located at Ravenshaw College, Cuttack. I could realise the problems faced by students who join the degree classes directly to make geology as their career. Further, shortage of teachers and supporting staff makes the problem more acute. I took up the book writing work after completing doctoral and post-doctoral research works.

Like other science subjects, both theory and practical are equally important in geology in addition to the field study. A few textbooks and a number of reference books on the theory part of the subject are available in Indian market. Realising the difficulties faced by the students in practical classes, I started writing a textbook of practical geology keeping the revised curriculum of the University Grants Commission in view.

The chemical analysis is necessary to determine the elemental composition of minerals and rocks. It is indispensable to determine the mineral formula, i.e. to identify the mineral from its composition and generate chemical data of rocks and minerals for different studies. This book deals with qualitative analysis of minerals as well as quantitative analysis of rocks and economic minerals. Since coal is regarded as an economic material at par with rocks, its proximate and ultimate analyses have been included in this book. In recent time more attention has been paid to the quality of water depending upon its field of use. So, a chapter on chemical analysis of water sample has been added. Many of the analyses included in this book are conventional tests and some can be performed with cost-effective flame- and spectro-photometers, ion-meter and water analyzer kit.

In the process of writing, a number of books written by different authors has been followed including incorporation of handouts provided to the students in practical classes. Any part of the text, method, statement and illustrations incorporated in the writing of this book is acknowledged. I am indebted and thankful to the authors and publishers of those referred materials. I am very much thankful to Prof. Wataru Maejima of Osaka City University, Japan, who taught me the methods of drawing/redrawing of figures, which made me independent to write this book to certain extent. I express my hearty gratitude to Prof. HK Sahoo of PG Department of Geology, Utkal University, who read major part of the manuscript, gave valuable suggestions to make the book complete in all respect and kindly agreed to write the 'Foreword' to this book. However, I accept the responsibility of any mistake, which might have been there inadvertently. In spite of critical scrutiny, some typographical errors might have been left out. I shall be very much thankful, if, lapses of any kind are communicated to me (rnhota@yahoo.com).

I express my gratitude to Prof. (Mrs) M Das, Dr PP Singh, Dr BK Ratha, Mr MR Mohapatra and Mr G Das of the PG Department of Geology, Utkal University for their encouragement and help during preparation of the manuscript. I express my sincere thanks to Dr D Satapthy and Dr SP Das of IMMT, Bhubaneswar and Prof. P Mohanty of PG Department of Chemistry, Utkal University for their discussions on different aspects of geochemical analysis.

I am thankful to Mr YN Arjuna, Senior Director, Publishing, Editorial and Publicity and M/s CBS Publishers & Distributors, New Delhi for their keen interest in publishing the book.

Last but not the least; I thank my wife Anjali and children Swayam Prakash and Soumya Ranjan for their inspiration and cooperation during preparation of the manuscript.

Rabindra Nath Hota

Contents

Introduction

INTRODUCTION

Chemical analysis is an integral part of mineralogy and petrology. Determination of water quality as well as environmental monitoring and assessment depend a lot on the chemical analysis of water, soil, solid and liquid wastes. The concentration of an element varies widely from sample to sample and can be expressed in terms of percent, ppm (parts per million) and ppb (parts per billion). The analytical procedures vary depending on the concentration of elements. When the concentration of an element is appreciably high, its qualitative and quantitative estimation can be made by conventional techniques available in a standard chemical laboratory. However, if the concentration is in micro-scale (ppm/ppb), sophisticated instruments like AAS (atomic absorption spectrometer), XRF (X-ray fluorescence spectrometer), ICP-MS (inductively coupled plasma mass spectrometer), etc. are necessary. Most of the procedures mentioned in this book are for the former category and can be followed and worked out in many college and university laboratories, particularly for analysis of minerals and rocks in which the concentration of constituent elements/oxides are appreciably high. Since the amounts of Na_2O and K_2O cannot be satisfactorily determined by volumetric analyses, the procedures of estimations of these two oxides by flame photometer have been included.

CLASSIFICATION OF GEOCHEMICAL ANALYSIS

The geochemical analysis has been divided into seven sections:

i. *Qualitative identification of elements present in minerals*: This section comprises a number of standard tests by which the presence of different constituent elements (or radicals) are inferred and confirmed leading to the identification of the mineral under consideration. Most of the minerals dealt within this section have economic connotation.

ii. *Quantitative analysis of minerals and rocks*: This section deals with different tests by which the constituent oxides of minerals and rocks can be estimated quantitatively. These are necessary for determination of chemical formulae of minerals and generation of data for computation of norm, construction of variation, triangular, discrimination and other diagrams in case of rocks.

iii. *Quantitative analysis of ores and minerals*: The commercial viability of ores and economic minerals depend largely on the amount of metal and useful ingredients they contain. A series of tests are given in this section by which quantitative estimation of metals and valuable ingredients can be made.

iv. Analysis of coal: The importance and use of coal can not be underestimated in the present day scene. It is extensively used as fuel starting from small-scale industries to large thermal power plants. The quality of coal depends on the quantity of its chemical constituents. The methods of proximate and ultimate analyses are given in this section.

v. Analysis of water sample: Determination of water quality is a pre-requisite for its use in different fields. This section deals with the analytical methods of surface and ground water from geological view-point. Since many constituent anions and cations are present in very minute quantities, the laboratory needs to be equipped with instruments like flame photometer, spectrophotometer, atomic absorption spectrometer, etc. Worked out problems and plottings of different parameters in standard diagrams have been given where ever possible.

vi. Analysis of soil sample: Soil is the weathered product of rock. It is necessary for growth of vegetation, which provides food to the animal kingdom. The type and quality of plant growth depend largely on the type of soil, which in turn is controlled by the physical and chemical characteristics of soil. This section deals with the analytical methods for determination of some important physical and chemical parameters of soil.

vii. Analysis of air sample: Certain constituents of air are expelled from volcanoes, ooze out from hot springs and produced by decay of organisms. These constituents, when exceed the permissible limits, contribute to environmental degradation. In this section, attempt has been made to estimate the amounts of suspended particulate matter as well as nitrogen and sulphur dioxides.

Qualitative Chemical Analysis of Minerals

Rocks consist of minerals, most of which are formed by inorganic processes of nature. A native mineral is composed of a single element and thus has only one radical. On the other hand, most of the minerals are chemical compounds having both basic and acid radicals. A broad classification of common minerals on the basis of chemical composition is given in Table 2.1.

Table 2.1: Classification of common minerals on the basis of chemical composition

Group/mineral	Chemical formula	Mineral	Chemical formula
Native metals			
Copper	Cu	Gold	Au
Platinum	Pt	Silver	Ag
Native semimetals			
Arsenic	As	Bismuth	Bi
Antimony	Sb		
Native nonmetals			
Diamond	C	Graphite	C
Sulphur	S		
Silicates			
Actinolite	$Ca_2(Mg,Fe)_5Si_8O_{22}(OH)_2$		
Aegirine	$Na(Al,Fe^{3+})Si_2O_6$	Albite	$NaAlSi_3O_8$
Almandine	$Fe_3Al_2Si_3O_{12}$	Andalusite	Al_2SiO_5
Analcite	$NaAlSi_2O_6 \cdot H_2O$	Andradite	$Ca_3Fe_2Si_3O_{12}$
Annite	$KFe_3(AlSi_3O_{10})(OH)_2$	Anorthite	$CaAl_2Si_2O_8$
Anthophyllite	$(Mg,Fe)_7Si_8O_{22}(OH)_2$	Antigorite	$Mg_6Si_4O_{10}(OH)_8$
Apophyllite	$KCa_4Si_8O_{20}F \cdot 8H_2O$		
Augite	$(Ca,Mg,Fe,Na)(Mg,Fe,Al)Si_2O_6$		
Beryl	$Be_3Al_2Si_6O_{18}$	Biotite	$K(Mg,Fe)_3(AlSi_3O_{10})(OH)_2$
Chabazite	$CaAl_2Si_4O_{12} \cdot 6H_2O$	Chamosite	$(Fe_5,Al)(AlSi_3)O_{10}(OH)_8$
Chlorite	$(Mg,Fe,Al)_6(Si,Al)_4O_{10}(OH)_8$		
Chloritoid	$(Fe,Mg)Al_2SiO_5(OH)_2$	Chondrodite	$Mg_5(SiO_4)_2(OH,F)_2$
Chrysotile	$Mg_6Si_4O_{10}(OH)_8$	Clinochlore	$(Mg_5,Al)(AlSi_3)O_{10}(OH)_8$
Clinohumite	$Mg_9(SiO_4)_4(OH,F)_2$	Clinozoisite	$Ca_2Al_3Si_3O_{12}(OH)$
Coesite	SiO_2	Cordierite	$(Mg,Fe)_2Al_4Si_5O_{18}$

(Contd...)

Table 2.1: Classification of common minerals on the basis of chemical composition *(Contd...)*

Group/mineral	Chemical formula	Mineral	Chemical formula
Cristobalite	SiO_2	Cummingtonite	$(Mg,Fe)_7Si_8O_{22}(OH)_2$
Diopside	$CaMgSi_2O_6$	Enstatite	$Mg_2Si_2O_6$
Epidote	$Ca_2(Al,Fe)_3Si_3O_{12}(OH)$	Fayalite	Fe_2SiO_4
Ferrisilite	$Fe_2Si_2O_6$	Forsterite	Mg_2SiO_4
Gedrite	$(Mg,Fe,Al)_7(Al,Si)_8O_{22}(OH)_2$		
Glaucophane	$Na_2Mg_3Al_2Si_8O_{22}(OH)_2$		
Grossular	$Ca_3Al_2Si_3O_{12}$	Grunerite	$Fe_7Si_8O_{22}(OH)_2$
Hedenbergite	$CaFeSi_2O_6$	Heulandite	$CaAl_2Si_7O_{18}\cdot6H_2O$
Hornblende	$(K,Na)_{0-1}(Ca,Na,Fe,Mg)_2$ $(Mg,Fe,Al)_5(Si,Al)_8O_{22}(OH)_2$		
Hypersthene	$(Mg,Fe)_2Si_2O_6$	Jadeite	$NaAlSi_2O_6$
Kaolinite	$Al_2Si_2O_5(OH)_4$	Kyanite	Al_2SiO_5
Lawsonite	$CaAl_2Si_2O_7(OH)_2\cdot H_2O$	Lepidolite	$K(Li,Al)_{2-3}(AlSi_3O_{10})(OH)_2$
Leucite	$KAlSi_2O_6$	Lizardite	$Mg_6Si_4O_{10}(OH)_8$
Margarite	$CaAl_2(Al_2Si_2O_{10})(OH)_2$	Marialite	$Na_4(AlSi_3O_8)_3Cl$
Meionite	$Ca_4(Al_2Si_2O_8)_3(CO_3,SO_4)$	Microcline	$KAlSi_3O_8$
Monticellite	$CaMgSiO_4$	Muscovite	$KAl_2(AlSi_3O_{10})(OH)_2$
Natrolite	$Na_2Al_2Si_3O_{10}\cdot2H_2O$	Nepheline	$(Na,K)AlSiO_4$
Nimite	$(Ni_5,Al)(AlSi_3)O_{10}(OH)_8$	Norbergite	$Mg_3SiO_4(OH,F)_2$
Omphacite	$(Ca,Na)(Fe,Mg,Al)(Si,Al)_2O_6$		
Orthoclase	$KAlSi_3O_8$	Pectolite	$NaCa_2(SiO_3)_3H$
Phlogopite	$KMg_3(AlSi_3O_{10})(OH)_2$		
Pigeonite	$(Ca,Mg,Fe)_2Si_2O_6$	Prehnite	$Ca_2Al(AlSi_3O_{10})(OH)_2$
Pyrope	$Mg_3Al_2Si_3O_{12}$	Pyrophyllite	$Al_2Si_4O_{10}(OH)_2$
Quartz	SiO_2	Rhodonite	$MnSiO_3$
Riebeckite	$Na_2(Fe,Mg)_3(Fe,Al)_2$ $Si_8O_{22}(OH)_2$		
Sanidine	$(K,Na)AlSi_3O_8$	Sillimanite	Al_2SiO_5
Sodalite	$Na_3Al_3Si_3O_{12}\cdot NaCl$	Spessartite	$Mn_3Al_2Si_3O_{12}$
Sphene	$CaTiSiO_5$	Spondumen	$LiAlSi_2O_6$
Staurolite	$Fe_2Al_9Si_4O_{23}(OH)$	Stilbite	$CaAl_2Si_7O_{18}\cdot7H_2O$
Stishovite	SiO_2	Talc	$Mg_3Si_4O_{10}(OH)_2$
Tephroite	Mn_2SiO_4	Topaz	$Al_2SiO_4(F,OH)_2$
Tourmaline	$(Na,Ca)(Fe,Mg,Al,Li)$ $Al_6(BO_3)_3Si_6O_{18}(OH)_4$		
Tremolite	$Ca_2Mg_5Si_8O_{22}(OH)_2$	Tridymite	SiO_2
Uvarovite	$Ca_3Cr_2Si_3O_{12}$		
Vesuvianite (idocrase)	$Ca_{10}(Mg,Fe)_2Al_4Si_9O_{34}(OH)_4$		
Wollastonite	$CaSiO_3$	Zircon	$ZrSiO_4$

Sulphides			
Argentite	Ag_2S	Arsenopyrite	$FeAsS$
Bornite	Cu_5FeS_4	Chalcocite	Cu_2S

(Contd...)

Table 2.1: Classification of common minerals on the basis of chemical composition *(Contd...)*

Group/mineral	Chemical formula	Mineral	Chemical formula
Chalcopyrite	$CuFeS_2$	Cinnabar	HgS
Cobaltite	$(Co,Fe)AsS$	Covellite	CuS
Energite	Cu_3AsS_4	Galena	PbS
Marcasite	FeS_2	Millerite	NiS
Molybdenite	MoS_2	Niccolite	$NiAS$
Orpiment	As_2S_3	Pentlandite	$(Ni,Fe)_9S_8$
Pyrargyrite	Ag_3SbS_3	Pyrite	FeS_2
Pyrrhotite	$Fe_{1-x}S$	Realgar	AsS
Sphalerite	ZnS	Stibnite	Sb_2S_3
Tetrahedrite	$Cu_{12}Sb_4S_{13}$	Wurtzite	ZnS
Halides			
Atacamite	$Cu_2Cl(OH)_3$	Chlorargerite	$AgCl$
Cryolite	Na_3AlF_6	Fluorite	CaF_2
Halite	$NaCl$	Sylvite	KCl
Oxides			
Cassiterite	SnO_2	Chromite	$FeCr_2O_4$
Chrysoberyl	$BeAl_2O_4$	Columbite	$(Fe,Mn)Nb_2O_6$
Corundum	Al_2O_3	Cuprite	Cu_2O
Franklinite	$ZnFe_2O_4$	Hematite	Fe_2O_3
Ilmenite	$FeTiO_3$	Magnetite	Fe_3O_4
Periclase	MgO	Perovskite	$CaTiO_3$
Pyrolusite	MnO_2	Rutile	TiO_2
Spinel	$MgAl_2O_4$	Tantalite	$(Fe,Mn)Ta_2O_6$
Thorianite	ThO_2	Uraninite	UO_2
Zincite	ZnO		
Hydroxides			
Brucite	$Mg(OH)_2$	Diaspore	$AlO(OH)$
Gibbsite	$Al(OH)_3$	Goethite	$FeO(OH)$
Manganite	$MnO(OH)$	Psilomelane	$BaMn_9O_{16}(OH)_4$
Carbonates			
Ankerite	$CaFe(CO_3)_2$	Aragonite	$CaCO_3$
Azurite	$Cu_3(CO_3)_2(OH)_2$	Calcite	$CaCO_3$
Cerussite	$PbCO_3$	Dolomite	$CaMg(CO_3)_2$
Kutnahorite	$CaMn(CO_3)_2$	Magnesite	$MgCO_3$
Malachite	$Cu_2CO_3(OH)_2$	Rhodochrosite	$MnCO_3$
Siderite	$Fe\ CO_3$	Smithsonite	$Zn\ CO_3$
Strontianite	$SrCO_3$	Witherite	$BaCO_3$
Nitrate			
Niter	KNO_3	Nitrite	$NaNO_3$
Borates			
Boracite	$Mg_3ClB_7O_{13}$	Borax	$Na_2B_4O_5(OH)_4\cdot8H_2O$
Colemanite	$CaB_3O_4(OH)_3\cdot H_2O$	Dumortierite	$Al_{6\frac{1}{2}-7}BSi_3O_{15}(O,OH)_3$

(Contd...)

Table 2.1: Classification of common minerals on the basis of chemical composition *(Contd...)*

Group/mineral	Chemical formula	Mineral	Chemical formula
Kernite	$Na_2B_4O_6(OH)_2 \cdot 3H_2O$	Sinhalite	$MgAlBO_4$
Ulexite	$NaCaB_5O_6(OH)_6 \cdot 5H_2O$		
Sulphates			
Anglesite	$PbSO_4$	Anhydrite	$CaSO_4$
Barite	$BaSO_4$	Celestite	$SrSO_4$
Hydrous sulphates			
Alunite	$KAl_3(SO_4)_2(OH)_6$	Antlerite	$Cu_3SO_4(OH)_4$
Chalcanthite	$CuSO_4 \cdot 5H_2O$	Epsomite	$MgSO_4 \cdot 7H_2O$
Gypsum	$CaSO_4 \cdot 2H_2O$		
Tungstates			
Scheelite	$CaWO_4$	Wolframite	$(Fe,Mn)WO_4$
Molybdate			
Wulfenite	$PbMoO_4$		
Chromate			
Crocoite	$PbCrO_4$		
Phosphates			
Amblygonite	$LiAl(PO_4)F$	Apatite	$Ca_5(PO_4)_3(OH,F,Cl)$
Autunite	$Ca(UO_2)_2(PO_4)_2 \cdot 8H_2O$	Lazulite	$(Mg,Fe)Al_2(PO_4)_2(OH)_2$
Monazite	$(Ce,La,Th,Y)PO_4$	Pyromorphite	$Pb_5(PO_4)_3Cl$
Torquise	$CuAl_6(PO_4)_4(OH)_8 \cdot 4H_2O$	Triphylite	$Li(Fe,Mn)PO_4$
Wavellite	$Al_3(PO_4)_2(OH)_3 \cdot 5H_2O$		
Vanadates			
Carnotite	$K_2(UO_2)_2(VO_4)_2$	Vanadinite	$Pb_5(VO_4)_3Cl \cdot 3H_2O$
Arsenates			
Erythrite	$Co_3(AsO_4)_2 \cdot 8H_2O$	Scorodite	$FeAsO_4 \cdot 2H_2O$

The mineral whose chemical analysis is to be done is grinded and powdered to less than 200 mesh in a porcelain mortar and pestle. In case of hard minerals agate mortar and pestle is used. The first step of identification is the qualitative chemical analysis that helps to determine the basic and acid radicals present in the minerals. In many cases the exact mineral can be conclusively identified. But polymorphous minerals like pyrite (FeS_2) and marcasite (FeS_2); andalusite (Al_2SiO_5), kyanite (Al_2SiO_5) and sillimanite (Al_2SiO_5) pose problem for proper identification. In these cases, the chemical analysis results are to be considered in conjunction with physical and/or optical properties of the minerals. Most of the chemical tests carried out for identification of minerals are dry tests, which suggest the presence of one or more elements. Wet tests are carried out for confirmation of the presence of basic radicals and identification of acid radicals. The tests are performed in the following order: (a) Closed tube test, (b) open tube test, (c) heating in charcoal cavity in oxidizing flame, (d) heating in charcoal cavity in reducing flame, (e) flame test, (f) microcosmic bead test, and (g) borax bead test. These tests are followed by wet confirmatory tests for basic radical and identification of the acid radical. All the dry tests should be performed, even if the reactions in some cases are negative, because the tests aim at proving with certainty, the presence of some element(s) and absence of others.

2.1. DRY TESTS FOR BASIC RADICALS

In these tests, the mineral powder is necessarily treated in the solid state to infer the presence of one or more basic radicals, though in a few cases it is moistened with a drop of water or wet chemical. Different dry tests are closed and open tube tests, heating in charcoal cavity in oxidizing and reducing flames, flame test, microcosmic and borax bead tests.

2.1.1. Closed Tube Test

Experiment	Observation	Inference
1. Take the mineral powder in a closed tube and heat strongly.	Orange sublimate	Presence of S
	Black sublimate with garlic smell	Presence of As
	Dark red volatile liquid when hot that turns into red or yellow solid with garlic smell when cold	Presence of As
	Red to dark yellow volatile liquid when hot that turns into crystalline sublimate with pungent smell when cold	Presence of Fe, Cu or Zn
	Sublimate is black when hot and brownish red when cold	Presence of Sb
	Black sublimate, red on rubbing	Presence of Hg
	Pale yellow to colourless liquid, white residue	Presence of Te
2. Heat the mineral powder in a closed tube with Na_2CO_3 and charcoal powder	Black mirror with garlic smell	Presence of As
	Globules of Hg are formed	Presence of Hg

2.1.2. Open Tube Test

Experiment	Observation	Inference
Take the mineral powder in an open tube and heat gently.	Sulphurous fumes yellow when hot and white when cold	Presence of S
	White crystalline sublimate volatile away with garlic smell	Presence of As
	Sublimate yellow when hot turns white when cold	Presence of Sb
	Gray globules	Presence of Hg
	Yellow sulphurous fumes when hot turns white when cold	Presence of Mo
	White sublimate fusible to colourless drops	Presence of Te
	White volatile sublimate	Presence of Se

2.1.3. Heating in Charcoal Cavity in Oxidizing Flame

	Experiment	Observation	Inference
1.	Take the mineral powder in a charcoal cavity, moisten with a drop of water and heat in oxidizing flame	Volatile sublimate that yellow near the cavity and white away from the cavity, garlic smell	Presence of As
		White sublimate near the cavity and bluish away from the cavity	Presence of Sb
		Incandescent mass, yellow when hot and white when cold	Presence of Zn
		Black magnetic residue	Presence of Fe
		Dark yellow when hot, yellow when cold	Presence of Pb
		Sublimate yellow when hot and colourless when cold	Presence of Mo
		Sublimate yellow when hot and white when cold	Presence of Sn
		Bright red when hot, paler when cold	Presence of Bi
		The residue added to water turns red litmus blue	Presence of Ca/Mg/Sr/ Ba/Be, etc.
		Sulphurous smell	Sulphide mineral
		Black to reddish brown residue, often iridescent	Presence of Cd
2.	Take the mineral powder in a charcoal cavity with potassium iodide and sulphur powder and heat in oxidizing flame Scarlet	Brilliant yellow residue	Presence of Pb
		Greenish-yellow residue and greenish yellow fumes	Presence of Bi
			Presence of Hg
3.	Moisten the mineral powder with CoNO₃ solution and heat in oxidizing flame	Grass green residue	Presence of Zn
		Bluish green residue	Presence of Sn
		Blue residue	Presence of Al
		Pink residue	Presence of Mg
		Dirty green residue	Presence of Sb
		Blue and glassy looking	Fusible silicates

2.1.4. Heating in Charcoal Cavity in Reducing Flame

	Experiment	Observation	Inference
1.	Take a mixture of the mineral powder and equal amount of $Na_2CO_3 + K_2CO_3$ in a charcoal cavity and heat in reducing flame	Strongly magnetic residue	Presence of Fe
		Feebly magnetic residue	Presence of Ni or Co
		Red spongy mass	Presence of Cu
		Silver white malleable residue	Presence of Ag
		Soft malleable residue that marks paper	Presence of Pb
		Tin-white, soft and malleable not marking paper	Presence of Sn
		Yellow bead, soft and malleable	Presence of Au

Experiment	Observation	Inference
2. Heat the mixture of mineral powder, charcoal powder and Na_2CO_3 in a charcoal cavity in reducing flame. Boil the residue with dil. HCl and tin granules	Violet coloured solution	Presence of Ti
	Blue coloured solution	Presence of W
	Opaque yellow solution	Presence of Cr
	Opaque bluish-green solution	Presence of Mn
	Insoluble mass	Presence of Si

2.1.5. Flame Test

Experiment	Colour of flame	Inference
Moisten the tip of a clean platinum wire with conc. HCl, dip it in mineral powder taken in a watch glass and heat the tip of the wire in oxidizing flame	Crimson	Presence of Sr
	Orange red	Presence of Ca
	Green flame with blue center	Presence of Cu
	Yellowish green	Presence of Ba
	Pale blue	Presence of Pb
	Golden yellow	Presence of Na
	Pale violet	Presence of K
	Carmine red	Presence of Li
	Green	Presence of Mo
	Greenish blue	Presence of Sb
	Pale green	Presence of Zn
	Bright green	Presence of Tl
	Yellowish green	Presence of B
	Bluish green	Presence of PO_4
	Whitish blue	Presence of As
	Azure blue	Presence of Se

2.1.6. Microcosmic Bead Test

Experiment	Colour of oxidizing flame	Colour of reducing flame	Inference
Make a bead of micro-cosmic salt [$Na(NH_4)$ $HPO_4 \cdot 4H_2O$] at the tip of a looped platinum wire, dip it in mineral powder and heat in oxidizing and reducing flames	Blue	Opaque red	Presence of Cu
	Red/green	Green	Presence of Cr
	Violet	Colourless	Presence of Mn
	Brownish red	Reddish	Presence of Fe
	Bright green	Green	Presence of Mo
	Colourless	Yellow/violet	Presence of Ti
	Colourless	Bluish green	Presence of W
	Blue	Blue	Presence of Co
	Yellow	Reddish yellow	Presence of Ni
	Greenish yellow	Fine green	Presence of U
	Skeletal residue	—	Presence of Si

2.1.7. Borax Bead Test

Experiment	Colour of bead in				Inference
	Oxidizing flame		Reducing flame		
	Hot	Cold	Hot	Cold	
Make a bead of borax at the tip of a looped	Green	Yellowish green	Green	Green	Presence of Cr
platinum wire, dip it in mineral powder	Yellow	Yellow/ colourless	Green	Green	Presence of Fe
and heat in oxidizing and reducing flames	Blue	Greenish blue	Blue	Dull red/ opaque	Presence of Cu
	Pink	Pink	Colourless	Colourless	Presence of Mn
	Yellow	Colourless/ white	Gray	Brown to violet	Presence of Ti
	Blue	Blue	Deep blue	Blue	Presence of Co
	Brown	Violet	Gray	Gray	Presence of Ni
	Yellow	Brown	Pale green	Green/ colourless	Presence of U
	Yellow	Colourless	Gray	Gray	Presence of Pb
	Yellow	Colourless	Gray	Gray	Presence of Zn

2.2. WET CONFIRMATORY TESTS FOR BASIC RADICALS

In these tests, solution of the mineral powder is treated with different solid salts as well as acid and alkali solutions. In some cases, H_2S gas is passed through the mineral solution. The confirmatory tests for different elements are as follows:

Experiment	Observation	Inference
1. (i) Dissolve the mineral powder in HNO_3. Add dil. HCl drop by drop till excess	White precipitate is formed	Presence of Ag is confirmed
(ii) Wash the precipitate in distilled water, add dil. NH_4OH drop by drop till excess and shake well	The precipitate dissolves and reappears with addition of dil. HNO_3	Presence of Ag is confirmed
(iii) Add K_2CrO_4 and HNO_3 to the salt solution	Brick-red precipitate is formed	Presence of Ag is confirmed
2. (i) Dissolve the mineral powder in HCl and add $NH_4(OH)$ solution in excess	Gelatinous white precipitate is formed	Presence of Al is confirmed
(ii) Add $Na(OH)$ solution to the mineral powder solution drop by drop till excess	White precipitate is formed, which is dissolved in excess $Na(OH)$	Presence of Al is confirmed

(Contd...)

Experiment	Observation	Inference
3. (i) Dissolve the mineral powder in HNO_3. Add about 0.5 ml of HCl to the solution and boil. Pass H_2S through the solution	Yellow precipitate is formed	Presence of As is confirmed
(ii) Dissolve the precipitate in acetic acid and add K_2CrO_4 solution	Yellow precipitate is formed	Presence of Ba is confirmed
4. (i) Dissolve the mineral powder in conc. HCl. Add solid NH_4Cl till saturation followed by dil. $NH_4(OH)$ till alkaline. Then add $(NH_4)_2CO_3$ solution	White precipitate is formed, which is soluble in water	Presence of Ca is confirmed
(ii) Dissolve the precipitate in acetic acid and add ammonium oxalate solution followed by $NH_4(OH)$ solution	White precipitate is formed	Presence of Ca is confirmed
5. (i) Prepare the mineral powder solution in HCl. Add solid NH_4Cl till saturation followed by dil. NH_4OH till alkaline. Pass H_2S through the mixture	Black precipitate is formed. (Original solution is pink)	Presence of Co is confirmed
(ii) Add KCl followed by solid sodium nitrite and dil acetic acid to the mineral powder solution	Yellow precipitate is formed	Presence of Co is confirmed
6. (i) Prepare the mineral powder solution in HCl. Add solid NH_4Cl till saturation followed by dil NH_4OH drop by drop till excess	Gelatinous bluish green precipitate is formed	Presence of Cr is confirmed
(ii) Heat a mixture of mineral powder with solid $Na_2(CO_3)$ and KNO_3 strongly on a platinum loop	Yellow bead is obtained	Presence of Cr is confirmed
(iii) Prepare a solution of the bead obtained above in water. Acidify the solution with acetic acid and add lead acetate solution	Yellow precipitate is formed	Presence of Cr is confirmed
7. (i) Dissolve the mineral powder in conc. HCl. Add $NH_4(OH)$ solution drop by drop	Initially bluish white precipitate is formed that dissolves to form deep blue solution with excess $NH_4(OH)$	Presence of Cu is confirmed
8. (i) Dissolve the mineral powder in conc. HCl and oxidise it with HNO_3. Add KCN solution	Brown red solution gives blue precipitate	Presence of Fe is confirmed
(ii) Dissolve the mineral powder in conc. HCl and oxidise it with HNO_3. Add ammonium sulpho-cyanide solution	Blood red colour appears	Presence of Fe is confirmed

(Contd...)

Experiment	Observation	Inference
(iii) Dissolve the mineral powder in conc. HCl and add $NH_4(OH)$ solution	Reddish brown flocculent precipitate is formed	Presence of Fe is confirmed
9. (i) Dissolve the mineral powder in HNO_3. Add dil. HCl drop by drop till excess	White precipitate is formed	Presence of Hg is confirmed
(ii) Add $SnCl_2$ solution to the mineral powder solution in HNO_3 drop by drop till excess	White precipitate is formed that turns gray with excess $SnCl_2$	Presence of Hg is confirmed
10. (i) Dissolve the mineral powder in HCl and add chlorplatinic acid	Yellow precipitate is formed	Presence of K is confirmed
(ii) Add two drops of cobalt nitrate solution followed by solid sodium nitrite and acetic acid to one ml of mineral powder solution	Yellow precipitate is formed	Presence of K is confirmed
11. (i) Prepare the mineral powder solution in HCl. Add solid NH_4Cl till saturation followed by dil. NH_4OH till alkaline. Finally disodium-hydrogen-phosphate solution is added	White precipitate is formed	Presence of Mg is confirmed
(ii) Take one drop of the solution in a test tube. Add 0.5 ml of dil. HCl followed by one drop of magneson reagent. Add NaOH solution in excess to make the solution alkaline	Blue precipitate is formed	Presence of Mg is confirmed
12. (i) Prepare the mineral powder solution in HCl. Add solid NH_4Cl till saturation followed by dil. NH_4OH till alkaline. Pass H_2S through the mixture	Buff (flesh) coloured precipitate is formed	Presence of Mn is confirmed
(ii) Heat a mixture of mineral powder with solid $Na_2(CO_3)$ and KNO_3 strongly on a platinum loop	Green bead is obtained	Presence of Mn is confirmed
(iii) Prepare a solution of the bead obtained above in water. Acidify the solution with H_2SO_4	Green solution turns pink	Presence of Mn is confirmed
13. Prepare the mineral powder solution in water and add potassium pyroantimonate solution	White precipitate is formed	Presence of Na is confirmed
14. (i) Prepare the mineral powder solution in HCl. Add solid NH_4Cl till saturation followed by dil. NH_4OH till alkaline. Pass H_2S through the mixture	Black precipitate is formed (original solution is green)	Presence of Ni is confirmed

(Contd...)

Experiment	Observation	Inference
(ii) Add dimethyleglyoxime reagent and dil. NH_4OH to the solution	Red precipitate is formed	Presence of Ni is confirmed
15. (i) Dissolve the mineral powder in dil HNO_3 and add dil. HCl drop by drop	White precipitate is formed	Presence of Pb is confirmed
(ii) Wash the precipitate with distilled water and boil	The precipitate dissolves on heating and reappears on cooling	Presence of Pb is confirmed
(iii) Add acetic acid followed by potassium chromate to the mineral solution	Yellow precipitate is formed	Presence of Pb is confirmed
16. (i) Dissolve the mineral powder in HCl. Add the solution to a beaker full of water	White precipitate is formed	Presence of Sb is confirmed
(ii) Heat the solution till boiling and pass H_2S slowly	Orange precipitate is formed	Presence of Sb is confirmed
17. (i) Dissolve the mineral powder in HNO_3. Add about 0.5 ml of HCl to the solution and boil. Pass H_2S through the solution	Chocolate brown or faint yellow precipitate is formed	Presence of Sn is confirmed
(ii) Add NaOH solution to the mineral powder solution in HNO_3, drop by drop till excess	White precipitate is formed that dissolves with excess NaOH	Presence of Sn is confirmed
18. (i) Dissolve the mineral powder in conc. HCl. Add solid NH_4Cl till saturation followed by dil. $NH_4(OH)$ till alkaline. Then add $(NH_4)_2CO_3$ solution	White precipitate is formed, which is soluble in water	Presence of Sr is confirmed
(ii) Dissolve the precipitate in acetic acid and add $CaSO_4$ solution	White precipitate is formed	Presence of Sr is confirmed
19. Dissolve the mineral powder in hot dil. H_2SO_4 and add tin foils	Blue solution is formed	Presence of Ta is confirmed
20. Dissolve the mineral powder in hot dil. H_2SO_4, dilute with cold water and add two drops of H_2O_2	Yellow solution is formed	Presence of Ti is confirmed
21. Dissolve the mineral powder in HCl and add zinc granules	Blue solution is formed	Presence of W is confirmed
22. (i) Dissolve the mineral powder in HCl, add a few drops of HNO_3 and filter. Add Na_2S to the filtrate	White precipitate is formed	Presence of Zn is confirmed
(ii) Add Na(OH) solution drop by drop till excess	White precipitate formed is dissolved in excess Na(OH)	Presence of Zn is confirmed

(Contd...)

Experiment	Observation	Inference
(iii) Take a mixture of the mineral powder, solid Na_2CO_3 and KNO_3 in platinum loop and heat strongly. Yellow bead is formed which is soluble in water producing yellow solution. Add Acetic acid and lead acetate solution	Yellow precipitate is formed	Presence of Zn is confirmed

2.3. TESTS FOR ACID RADICALS

After determination of basic radical, the acid radical of the mineral is determined by a set of wet tests. The tests are given below.

	Experiment	Observation	Inference
1.	Take 1 ml of dil. HCl in a test tube and heat till boiling. Add a pinch of mineral powder to the acid	Effervescence takes place with release of a colourless and odorless gas (CO_2) that turns lime water milky	May be due to carbonate (CO_3^{2-})
		Effervescence takes place with evolution of a colourless gas with rotten egg smell that turns a filter paper dipped in lead acetate solution black	May be due to sulphide (S^{2-})
2.	Take a pinch of mineral powder in a test tube, add few drops of conc. H_2SO_4 and warm the test tube gently	Effervescence takes place with evolution of a colourless gas that produces fumes in moist air. When a glass rod dipped in conc. NH_4OH is exposed to the gas, white fumes are produced.	May be due to chloride (Cl^-). Perform the tests as given in experiments 3 and 4 below
		White vapour with slightly brownish appearance is evolved	May be due to nitrate (NO_3^-). Perform the confirmatory tests as given in experiments 5 and 6 below
3.	Take a pinch of mineral powder in a test tube, add about 1 ml of distilled water and shake well. Acidify the solution with dil. HNO_3. If the powder does not dissolve in distilled water, prepare the solution in dil HNO_3. Add $AgNO_3$ solution till precipitation is complete	A curdy white precipitate is formed. Wash the precipitate with distilled water, add dil. NH_4OH solution and shake well. The precipitate dissolves	Presence of chloride ion (Cl^-) is confirmed

(Contd...)

	Experiment	Observation	Inference
4.	Heat a mixture of the mineral powder with MnO_2 and 0.5 ml of conc. H_2SO_4 gently in a test tube	Greenish-yellow gas is evolved	Presence of chloride ion (Cl^-) is confirmed
5.	Take a pinch of mineral powder in a test tube, add one ml of 50% H_2SO_4 and few pieces of copper turnings.	The solution turns green or bluish green with evolution of brown fumes. A filter paper dipped in freshly prepared $FeSO_4$ solution turns black when exposed to the fumes	Presence of nitrate ion (NO_3^-) is confirmed
6.	Take a pinch of mineral powder, add 0.5 ml of conc. H_2SO_4, cool under tap water and add freshly prepared $FeSO_4$ solution slowly	A brown ring is formed at the junction of two solutions	Presence of nitrate ion (NO_3^-) is confirmed
7.	Take a pinch of mineral powder and 0.5 ml of conc. H_2SO_4 in a test tube and heat	Greasy bubbles of HF is evolved that causes deposition of white film of silica on a drop of water held at the mouth of the tube	Presence of fluoride (F^-). Perform the test below
8.	Heat a mixture of mineral powder, glass powder and $KHSO_4$	A ring of white sublimate is formed	Presence of fluoride (F^-) is confirmed
9.	Prepare mineral powder solution in HCl in a test tube and add $BaCl_2$	White precipitate is formed, which is insoluble in HCl even on boiling	Presence of sulphate (SO_4^{2-}). Perform the test below
10.	Heat a mixture of mineral powder, charcoal powder and Na_2CO_3 on a charcoal block. Add the residue to HCl taken in a test tube and heat slightly	H_2S gas is evolved that turns a filter paper dipped in lead acetate black	Presence of sulphate (SO_4^{2-}) is confirmed
11.	Prepare mineral powder solution in HCl in a test tube. Take 0.5 ml of ammonium-molybdate and equal volume of conc. HNO_3 in a test tube. Shake the test tube well and warm up to 40° C. Add a few drops of mineral powder solution	Yellow precipitate is formed	Acid radical is phosphate (PO_4^{3-})
12.	Take a pinch of mineral powder in a test tube, add 2 ml of HCl and heat gently	Gelatinise silica skeleton is formed	Presence of silica (Si)
13.	Take a pinch of mineral powder in a test tube, add 2 ml of conc. H_2SO_4 and heat gently	A deep red solution is formed. The colour disappears if the solution is heated too much or cooled by adding water	Presence of tellurium (Te)

If the above tests fail, the acid radical may be oxide (O^{2-}) or hydroxide (OH^-).

Quantitative Analysis of Rocks and Minerals

The rocks are composed of different minerals. The constituent minerals are identified by their megascopic and/or microscopic characters and their amounts are quantitatively determined by modal analysis. Each mineral has its unique chemical composition. Like a mineral, a rock as a whole is also a chemical compound. The chemical composition of a rock is expressed in terms of its oxide constituents determined in the chemical laboratory by a set of procedures. Similar procedures are also applicable for minerals. The common methods followed in chemical analysis of rocks and minerals are as follows.

3.1. GRINDING OF SAMPLE

To facilitate complete chemical reactions, the rock or mineral is finely powered to size less than 200 mesh (0.074 mm). This is achieved by breaking the sample into smaller size by hammer followed by crushing, grinding and sieving. In many instances grinding is done manually by porcelain or agate mortar and pestle.

3.2. OPENING OF SAMPLE

This is the chemical transformation of the rock or mineral sample into a desired form so that chemical reactions are facilitated. There are two methods to open a sample, viz. wet method and dry method. In wet method, the sample is digested by an acid or a mixture of acids to convert the complex compounds into simpler forms. In many instances, heating accelerates the chemical action of acid(s). The dry method involves fusion of rock sample with one or more solid chemicals that result in the conversion of complex compounds into simpler forms. The fusion is carried out in refractory crucibles made of platinum, nickel, iron, zirconium, palladium, porcelain, silica etc. The selection of fusion mixture depends on the nature of mineral or rock sample and the type of the desired end product. To speed up the fusion process, a suitable flux is added to the mixture. The end product of the opening process is known as *stock solution*. Quantitative estimation of the oxide constituents is carried out by a set of chemical analyses for each of which a small amount of the stock solution (*aliquot*) is taken.

The above-mentioned preliminary treatments are also applicable for metallic ores and non-metallic minerals of economic value.

A rock or mineral is generally composed of some or all of the major oxides like SiO_2, Al_2O_3, Fe_2O_3, FeO, MgO, CaO, Na_2O, K_2O, TiO_2, MnO, P_2O_5, CO_2, SO_3, H_2O, minor oxides like Cr_2O_3, ZrO_2, NiO, BaO including elements like S, Cl and F. Some of these constituents like SiO_2, H_2O, CO_2, S and loss of ignition (LOI) are determined by gravimetric analyses while Al_2O_3, FeO, Fe_2O_3, MgO, CaO, TiO_2, MnO, Na_2O, K_2O, P_2O_5 and F are estimated by titrimetry or volumetric analyses. Most of the standard gravimetric and titrimetric (volumetric) procedures developed and adopted by the Indian Bureau of Mines, for quantitative estimation of the constituent oxides stated above other than Na_2O and K_2O and the titrimetric procedures of estimation of Na_2O and K_2O proposed by Bassett, et al. (1978) are given in this book. The solutions for these two oxides are to be prepared by the methods proposed by Shapiro and Brannock (1962). Since the titrimetric procedures for estimation of Na_2O and K_2O are tedious and do not provide satisfactory results always, the elemental concentration of Na and K are determined by flame photometer, from which the amounts of Na_2O and K_2O can be estimated.

3.3. ESTIMATION OF SiO_2 BY GRAVIMETRIC ANALYSIS

Silica is a major constituent of almost all the rocks and is also associated with different economic minerals in variable proportions. The procedures followed for quantitative estimation of silica in different rocks and minerals are different and depend on the chemical composition of the rock or mineral. The procedures followed by Indian Bureau of Mines for estimation of SiO_2 in aluminosilicates (common rocks) are outlined below.

3.3.1. Sodium Carbonate (Na_2CO_3) Fusion Method

This method is applicable when the rock contains insignificant amount of fluorine.

i. Take about 1 gram of mineral or rock powder in a platinum crucible of 20–30 ml capacity and thoroughly mix 4–6 g of anhydrous Na_2CO_3 with it.

ii. Heat the mixture gradually to a maximum temperature of 1000 °C. Maintain the temperature till the reaction is over and the mass is quiescent.

iii. Cool the crucible and keep it in a 250 ml beaker in horizontal position.

iv. Cover the crucible completely with demineralised water, add 10–15 ml of conc. HCl to the content; cover the beaker by a watch glass and allow it to remain as such till the reaction is complete.

v. Heat the beaker in a low flame for a few minutes and then transfer it onto a hot plate.

vi. Keep the beaker on the hot plate with low temperature (about 110°C) till the solvent evaporates completely. Stir the content intermittently to avoid formation of lumps. Keep the beaker on the hot plate for 1 hour.

vii. Remove the beaker from the hot plate and cool.

viii. Moisten the content with 1–2 ml of conc. HCl, add 50 ml of demineralised water and boil in a low flame till reaction is complete.

ix. Take the beaker out of burner, cool and filter through Whatman No. 40 filter paper. Preserve the filtrate (stock solution) for estimation of other oxides, if necessary.

x. Wash the residue 5–6 times with hot water and transfer into a platinum crucible along with the filter paper. Ignite the filter paper in a muffle furnace at 1000°C.

xi. Cool the mass in a desiccator and weigh. Let the weight is W_1.

xii. Moisten the mass with demineralised water and add 2–3 drops of dil. H_2SO_4 and 10–15 ml of 48% HF (density 1.15 g/ml).

xiii. Heat the crucible initially in a low flame (on asbestos sheet on a low flame or on a low temperature hot plate) and finally at 1000°C for about 5 minutes till all the liquid is expelled.

xiv. Cool the crucible in a desiccator and weigh the mass. Let it is W_2.

Calculation

$$SiO_2\% = \frac{W_1 - W_2}{W_1} \times 100$$

3.3.2. Borate Fusion Method (When the Sample Contains Appreciable Amount of F)

a. Mix 0.5 g of borax to the mixture at step (i) mentioned above in article 3.3.1.

b. Follow the steps (ii) to (iv) as mentioned in article 3.3.1.

c. Wash the watch glass into the beaker.

d. Heat the content slowly on low flame to remove the mass from the crucible.

e. Remove the crucible from the heating zone and wash thoroughly.

f. Transfer the content of the beaker into a China-clay dish of about 15–20 cm in diameter.

g. Heat the China-clay dish on a low flame to expel the liquid by evaporation.

h. Take the dish out from the flame and add 4–5 drops of dil. H_2SO_4 and 10 ml methanol.

i. Ignite the content in open flame with continuously stirring. Apple green flame will indicate the presence of borate.

j. Cool the dish and repeat the processes (h) and (i) with additional quantity of H_2SO_4 and methanol till all the borate is burnt.

k. Follow the steps (viii) to (xiv) mentioned above to determine the amount of silica (SiO_2).

3.4. ESTIMATION OF Al_2O_3 BY EDTA (DISODIUM DIHYDRATE SALT OF ETHYLENE-DIAMINE-TETRA-ACETIC ACID) COMPLEXOMETRY BACK TITRATION METHOD

i. Take 50 ml aliquot from the stock solution [3.3.1 (ix)] in a beaker and neutralise it with 10% NaOH solution till a tea-red colour appears. There should not be any precipitation of iron. If it happens, discard the solution and take a fresh one.

ii. Add the solution to 50–75 ml 10% NaOH solution (10 ml NaOH solution + 90 ml distilled water; 1 M NaOH solution is prepared by dissolving 40 g NaOH pellets in 1 liter distilled water) taken in a 250 ml beaker with constant stirring.

iii. Add few drops of bromine water and a pinch of Na_2CO_3 to the solution and boil the mixture on a burner.

iv. Cool the solution and filter through Whatman No. 41 filter paper.

v. Transfer the content into a 500 ml beaker and add a little filter paper pulp.

vi. Acidify the filtrate slowly with dilute HCl. Precipitation of $Al(OH)_3$ will take place.

vii. Dissolve the precipitate with dilute HCl.

viii. Add EDTA solution. [Normally 25 ml of 0.05 M EDTA (18.612 g of disodium dihydrate EDTA salt in 1 liter distilled water) is added for estimation of Al_2O_3 in bauxite, clay, kyanite, etc. and 10 ml. of 0.02 M EDTA (7.449 g of disodium dihydrate EDTA salt in 1 liter distilled water) is added for estimation of Al_2O_3 in iron and manganese ores, limestone, etc.]

ix. Boil the solution.

x. Cool the solution and adjust the pH to 5.5 by addition of HCl and NaOH (or NH_3).

xi. Add 50 ml 5.5 pH sodium acetate buffer.

xii. Add 5–6 drops of xylenol orange indicator. (The indicator will impart a golden yellow colour to the solution.)

xiii. Titrate the solution against 0.05 M or 0.02 M standard zinc acetate solution (10.975 or 4.39 gram of zinc acetate dihydrate in 1 liter distilled water) till the colour changes from yellow to orange red. (The colour should be observed on the rim of the titrating solution, which is a back titration technique.)

xiv. Determine zinc acetate dihydrate equivalent for the volume of standard EDTA salt added.

xv. Find out the amount of standard EDTA solution (or equivalent zinc acetate) consumed by aluminium ion for complexion.

xvi. Estimate Al_2O_3 by following equivalence:
1 ml 1 M EDTA = 1 ml 1 M Zn-acetate = 0.05098 g Al_2O_3

Note: Very often Zn interferes with this estimation. A modified method for estimation of zinc and aluminum is as follows:

a. Determine first equivalence point following the above-mentioned procedure. This indicates the total amount of Al_2O_3 and zinc.

b. Add 1–2 g of NaF to the solution after first titration and boil.

c. Cool the solution and titrate against standard zinc acetate dihydrate solution.

d. Colour changes from golden yellow to orange red.

e. Determine the volume of EDTA required for complexing Zn ions by the formula:

1 ml 1 M EDTA = 1 ml 1 M Zn-acetate = 0.06537 g Zn

3.5. ESTIMATION OF IRON (Fe, FeO and Fe_2O_3)

Iron can be quantitatively estimated by potassium dichromate and potassium permanganate methods.

3.5.1. Potassium Dichromate Method

i. Take 50 ml of aliquot from the main stock solution [(3.3.1 (ix)] in a 250 ml beaker.

ii. Add 2–3 drops dilute solution of methyl orange that imparts a very faint red colour to the solution.

iii. Add 2 g of NH_4Cl with constant stirring.

iv. Precipitate mixed oxide (R_2O_3) by dropwise addition of dilute NH_3 till the colour of indicator turns yellow. Add 2–3 drops in excess.

v. Boil the content on a burner.

vi. Cool the content and filter through Whatman No. 41 filter paper. Preserve the filtrate for estimation of CaO and MgO, if necessary.

vii. Wash the precipitate 5–6 times with hot water and transfer into the original beaker carefully with a jet of hot water.

viii. Dissolve the precipitate in conc. HCl. Wash the sides of the beaker.

ix. Heat the beaker to 80–90°C on burner (avoid boiling) and add 10% $SnCl_2$ solution (10 g $SnCl_2$ in 90 ml distilled water) dropwise with constant stirring to reduce iron. Yellow colour of $FeCl_3$ disappears due to conversion of ferric ion to ferrous state. Add 1–2 drops of $SnCl_2$ in excess.

x. Cool the solution in cold water.

xi. Add 10 ml $HgCl_2$ solution and stir. A silky white precipitate of mercurous chloride will appear. If the precipitate is black, discard the solution and repeat the process with a new aliquot. The black colour indicates the reduction of mercurous chloride to elemental mercury that adds inaccuracy to the estimation.

xii. Add 10–15 ml sulphuric–phosphoric acid mixture and dilute the solution to about 150 ml.

xiii. Add 2–3 drops of barium diphenylamine sulphonate indicator solution to the solution. Avoid use of excess of indicator as it imparts intense dirty green colour near the end point and confuses the actual end point.

xiv. Titrate the solution against standard 1N potassium dichromate solution (294.18 g $K_2Cr_2O_7$ in 1 liter distilled water) with constant stirring till a constant stable violet colour appears.

xv. Estimation of iron is made by following equivalence:

$$\text{1 ml 1N } K_2Cr_2O_7 = 0.05584 \text{ g Fe} = 0.07185 \text{ g FeO} = 0.07985 \text{ g } Fe_2O_3$$

In most of the rocks, iron is present both in ferrous (FeO) and ferric (Fe_2O_3) states. In the experiment described above, total iron as Fe_2O_3 is found out. In order to determine the exact amount of both the constituents, absolute amount of FeO is determined by potassium permanganate method given below. Amount of Fe_2O_3 is calculated by formula: percent of Fe_2O_3 = percent of total Fe (as Fe_2O_3) – (percent of FeO × 1.1113)

Gravimetric factors:

$$Fe \quad = 0.69944 \times Fe_2O_3$$
$$Fe \quad = 0.77731 \times FeO$$
$$FeO \quad = 0.89981 \times Fe_2O_3$$
$$Fe_2O_3 = 1.11130 \times FeO$$

The method of estimation of FeO is given on next page.

3.5.2. Potassium Permanganate Method

Reagent preparation

a. *Boric acid solution:* Dissolve 100 g boric acid in 1 litre hot distilled water with constant stirring. Cool the solution and dilute to make 2 litres.

b. *Potassium dichromate solution (0.1 N):* Take 9.810 g finely ground analar grade $K_2Cr_2O_7$ in a weighing bottle and dry at 110°C in an oven for more than 2 hours. Weigh the bottle with the contents and transfer the content into a 2-litre volumetric flask through a funnel. Weigh the bottle once again to know the true weight of $K_2Cr_2O_7$ used. Wash the funnel and neck of the flask. Dissolve the reagent and make 2 litres with distilled water. The equivalent weight of potassium dichromate is 49.035 g.

$$N = \frac{\text{Wt. of } K_2Cr_2O_7 \text{ in g}}{2 \times 49.035 \text{ g}}$$

Normality of potassium dichromate solution can be checked by using a standard Fe solution. 10 ml of Fe solution requires approximately 10 ml of 0.1 N $K_2Cr_2O_7$ (29.4 g $K_2Cr_2O_7$ in 1 liter distilled water).

c. *Standard iron solution:* Take about 1.3 g of pure iron wire or chips. Wash it by ether, dry in air and keep in a 150 ml beaker. Add 50 ml of 1:1 HCl (100 ml 36% HCl + 100 ml distilled water) to the iron, cover by a watch glass and warm on a steam bath till the iron completely goes into solution. Cool the solution, transfer to a 250 ml flask and make up to 250 ml volume.

The method of estimation of FeO

i. Take 0.5 g of 200-mesh size powdered sample in a 45 ml platinum crucible with a tight fitting lid and a 2 mm central hole.

ii. Add 1 ml of water and give a twisting motion to the crucible to spread out the sample evenly over the bottom.

iii. Add 2–3 drops of 1:1 H_2SO_4 (100 ml 98% concentrated H_2SO_4 + 100 ml distilled water) to decompose any carbonates if present. Allow the crucible to stand still until the reaction is over.

iv. Take 5 ml of 18 M H_2SO_4, 5 ml of concentrated HF and 10 ml of water in a 50 ml platinum dish. Heat the mixture slightly.

v. Heat the covered crucible with Fe solution by a low-flame Bunsen burner.

vi. Slip the cover of the crucible to one side quickly and carefully add the hot mixture from the platinum dish. Replace the cover of the crucible and heat on a Bunsen burner until the contents boil and steam comes out through the hole. Adjust the flame of the burner so that boiling goes on slowly. Continue heating for 10 minutes.

vii. Take 200 ml of water in a 600 ml beaker. Add 50 ml boric acid solution, 5 ml 18 M H_2SO_4 and 5 ml 85% H_3PO_4 to water and mix well.

viii. Hold the crucible firmly with the platinum-tipped tongs at the end of heating period. Press the cover firmly with a glass rod and quickly submerge the crucible beneath the acid solution in the 600 ml beaker. No part of the tong except the platinum tips should enter into the acid solution.

ix. Take out the cover of the crucible with a glass rod and stir the content slowly till all the soluble material is dissolved.

x. Titrate the solution immediately with 0.1N $KMnO_4$, (prepared previously by dissolving 3.3 g of $KMnO_4$ in 1 litre of water and allowed to stand for 3 days at room temperature) till first appearance of a permanent pink tinge, or with 0.1N $KMnO_4$ with 6 drops of 0.2 % barium diphenylamine sulfonate that serves as indicator till first appearance of a permanent blue-violet colour.

Computation:

$$FeO\ (\%) = \frac{7.185 \times \text{Titrant in ml} \times \text{Normality}}{\text{Sample wt. in g}}$$

3.6. ESTIMATION OF CaO

CaO can be determined by EDTA complexometry and oxalate methods.

3.6.1. EDTA (Disodium Dihydrate Salt of Ethylenediaminetetra-acetic Acid) Complexometry Method

i. Take about 50% of the filtrate obtained in iron estimation [3.5.1. (vi)] in a 250 ml beaker.

ii. Add 2–3 drops of triethanol amine to the filtrate and stir by a glass rod.

iii. Add a pinch of hydroxylamine hydrochloride or ascorbic acid, 50 ml of 20% KOH solution (20 g KOH + 80 ml distilled water) and 5–6 drops of Patton and Readers indicator to the solution. The indicator gives a rose red colour to the solution.

iv. Titrate the solution against standard 1M EDTA disodium dihydrate salt solution (372.24 g in 1 liter distilled water) till colour changes from rose red to blue.

$$1\ \text{ml of 1M EDTA} = 0.05608\ \text{g of CaO}$$

3.6.2. Oxalate Method

i. Take about 50% of the filtrate obtained in iron estimation [3.5.1. (vi)] in a 250 ml beaker.

ii. Acidify the solution with HCl. Calcium will be precipitated as calcium oxalate.

iii. Filter, wash and dissolve the precipitate in hot dilute H_2SO_4. Titrate the resulting solution against standard potassium permanganate solution to pink colour.

iv. Take the ammonical filtrate from the mixed oxide precipitation in a 500 ml beaker and acidify the solution with HCl using methyl red as indicator.

v. Add about 0.5 g of ammonium oxalate and stir well to make the solution ammonical and boil it for about 10 minutes.

vi. Remove the beaker from the flame and allow the precipitate to settle for about 1 hour.

vii. Filter the content through Whatman No. 40 filter paper and wash the residue by hot water till it is free from oxalate (test the filtrate with a drop of dilute permanganate solution and a drop of dilute H_2SO_4).

viii. Collect and wash the filtrate for estimation of MgO, if necessary.

ix. Wash the precipitate with water into a beaker.

x. Add about 25 ml of 1:1 H_2SO_4 (100 ml concentrated H_2SO_4 + 100 ml distilled water) and warm the solution up to 70 to 80°C.

xi. Titrate the solution with standard permanganate solution till a permanent pink colour appears.

xii. Calculate the percentage of CaO from the following equation.

$$CaO\ (\%) = \frac{2.8 \times A \times B}{C}$$

where,

A = Volume in ml of standard permanganate solution used

B = Normality of permanganate solution

C = Weight (in g) of the sample representing the aliquot taken

3.7. ESTIMATION OF MgO

MgO can be estimated by EDTA complexometry and pyrophosphate methods.

3.7.1. EDTA (Disodium Dihydrate Salt of Ethylenediaminetetra-acetic Acid) Complexometry Method

    i. Take about 50% of the filtrate obtained in iron estimation [3.5.1. (vi)] in a 250 ml beaker.

    ii. Add 2–3 drops of triethanol amine to the filtrate and stir by a glass rod.

    iii. Add a pinch of hydroxylamine hydrochloride or ascorbic acid, 50 ml of 10 pH ammonia-ammonium chloride buffer and 5–6 drops of Eriochrome Black T indicator to the solution. The indicator imparts a brilliant red colour to the solution.

    iv. Titrate the solution against standard 1M EDTA disodium dihydrate solution (372.24 g in 1 liter distilled water) till colour changes from red to blue. This titration estimates calcium and magnesium together. Magnesium equivalent of EDTA is found out by subtracting the volume necessary for CaO estimation.

$$1\ ml\ of\ 1M\ EDTA\ solution = 0.0409\ g\ of\ MgO$$

3.7.2. Pyrophosphate Method

    i. Take the filtrate from the precipitation of calcium oxalate [3.6.2. (viii)] in a 500 ml beaker.

    ii. Add 50 ml HNO_3 and 20 ml conc. HCl to the filtrate and boil the mixture vigorously to destroy the organic salt, if present.

    iii. When the material is nearly dry, add 10 ml conc. HCl and 100 ml water and boil once again.

    iv. If the solution is not clear, filter through Whatman No. 40 filter paper and discard the residue.

    v. Add bout 0.5 g of diammonium hydrogen phosphate [$(NH_4)_2HPO_4$] to the filtrate and stir well to dissolve the phosphate. Add a few drops of methyl red indicator.

    vi. Add conc. ammonia solution with continuous stirring until the indicator turns yellow. Continue stirring for 5 minutes more and finally add 5 ml of concentrated ammonia solution in excess. Magnesium ammonium phosphate hexahydrate ($MgNH_4PO_4.6H_2O$) will be precipitated at room temperature.

    vii. Allow the solution to stand in a cool place overnight and filter through No. 40 Whatman filter paper on the next day.

    viii. Wash the precipitate on the paper with cold 0.1% aqueous ammonia solution until it is free from chloride ions.

ix. Transfer the filter paper along with the precipitate into a platinum or porcelain crucible of known weight.

x. Burn the filter paper at a low heat and then ignite to a constant weight in a muffle furnace at 1000 to 1100°C. Magnesium ammonium phosphate hexahydrate will be converted to magnesium pyrophosphate ($Mg_2P_2O_7$).

xi. Cool the crucible in a desiccator and weigh the residue.

xii. Percentage of MgO is calculated by the formula given below:

$$MgO\ (\%) = \frac{3.623 \times A}{B}$$

where, A = weight of the $Mg_2P_2O_7$ in g

B = weight (in g) of the sample represented by aliquot

3.8. ESTIMATION OF Na₂O AND K₂O

Na_2O and K_2O in rock and mineral samples can be estimated by volumetric analyses, for which stock solution can be prepared by the following method.

3.8.1. Preparation of Stock Solution

Reagents: Acid mixture-1: 1 litre 48% HF, 165 ml conc. H_2SO_4 and 40 ml conc. HNO_3.

Acid mixture-2: 100 ml of 72% $HClO_4$ and 100 ml conc. HNO_3

i. Take 0.5 g of samples in two Teflon beakers.

ii. Add 15 ml acid mixture-1 to each beaker and swirl the beakers to wet the sample powder.

iii. Cover the beakers and place them on the steam bath for overnight heating.

iv. Take out the covers and heat the beakers on the steam bath till acid fumes cease to evolve.

v. Transfer the contents of two Teflon beakers to two Vycor beakers of 400 ml capacity.

vi. Heat the Vycor beakers on hot plate till SO_3 fumes come out.

vii. When SO_3 fumes stop evolving, add 4 drops of acid mixture -2 to the content of each beaker and heat again on hot plate. Continue heating until strong fumes come out and colour of the organic matter (if present) disappears.

viii. Remove the beakers from the hot plate, cool and add 225 ml distilled water, 5 ml conc. HNO_3 and 1 ml of 0.2% hydrazine sulphate solution to each beaker.

ix. Heat the beakers on a hot plate. If brown precipitate of MnO_2 remains after solutions have been boiled for few minutes, add 1 ml of 0.2% hydrazine sulphate solution to each beaker and heat the beakers for 30 minutes.

x. Take out the beakers from the hot plate and allow them to cool. A small amount of residue can be ignored. However, if appreciable amount of residue remains, it should be separated and identified by suitable methods.

xi. Cool the solutions to room temperature and transfer them into two 250 ml volumetric flasks, make them up to 250 ml and transfer to polythene bottles. These two solutions should be used for estimation of Na_2O and K_2O.

The stock solutions prepared above can be used for estimation of Na_2O and K_2O by titrimetry methods and/or by flame photometer.

3.8.2. Estimation of Na₂O by Titrimetry Method

The sodium of the stock solution is precipitated as sodium zinc uranyl acetate $[NaZn(UO_2)_3(CH_3COO)_3.6H_2O]$ and indirectly determined from the amount of Zn by titration with disodium dihydrate salt of EDTA.

Indicators

a. Grind 1 g of Solochrome Black (Erichrome Black T) with 10 g AR KNO_3.

b. 1 M $(NH_4)_2CO_3$ is prepared by dissolving 48 g of solid $(NH_4)_2CO_3$ in 500 ml deionised water.

Procedure

i. Treat about 1.5 ml of aliquot with 15 ml zinc uranyl acetate reagent.

ii. Stir vigorously for 30 minutes and allow to stand for about 1 hour.

iii. Filter through a weighted porcelain or sintered glass-filtering crucible of porosity 4.

iv. Wash the precipitate 4 times with 2 ml of precipitating reagent (zinc uranyl acetate) and 10 times with 95% ethanol saturated with sodium uranyl acetate at room temperature and finally with a little dry diethyl ether or acetone.

v. Dry for 30 minutes at 55–60°C.

vi. Weigh the sodium zinc uranyl acetate.

vii. Keep the filtered crucible inside a 400 ml pyrex beaker and add 5 ml of 1 M HCl (8.3 ml conc. HCl in 100 ml distilled water).

viii. After a few minutes, add 50 ml deionised water and boil.

ix. Allow the solution to cool.

x. Remove the crucible with tongs and wash carefully into the beaker.

xi. Neutralise the combined filtrate and washings (total about 100 ml) with 1 M $(NH_4)_2CO_3$ [96.09 g of $(NH_4)_2CO_3$ in 1 liter distilled water]. Add 2 ml buffer mixture (pH = 10) and 30 mg of Solochrome Black (Erichrome Black T) indicator.

xii. Titrate with standard 0.001 M EDTA disodium dihydrate solution (0.372 g of EDTA disodium dihydrate in 1 liter distilled water) till the colour changes from yellowish red to blue.

xiii. Calculate the amount of Na from the volume of ETDA required for complexing Zn ions.

$$1 \text{ ml } 1 \text{ M EDTA} = 1 \text{ ml } 1 \text{ M Zn-acetate} = 0.06537 \text{ g Zn}$$

3.8.3. Estimation of K₂O by Titrimetry Method

The potassium of the stock solution is precipitated with excess Na-tetraphylborate solution. The excess of the reagent left out from chemical reaction is determined by titration with $Hg(NO_3)_2$ in the presence of $Fe(NO_3)_3$ and Na-thiocyanate indicators.

Reagents

a. Na-tetraphylborate solution—add 6 g of Na-tetraphylborate and 1 g of $Al(OH)_3$ gel to 100 ml of distilled water. Shake well, filter through Whatman No. 40 filter paper. Add 15 ml of 0.1 M NaOH (4 g NaOH in 1 liter distilled water) to give a pH of 9. Make up to 1 litre and preserve in a polythene bottle.

b. $Hg(NO_3)_2$ solution—add 10 g of $Hg(NO_3)_2$ in 800 ml of distilled water and add 20 ml of HNO_3 and make up to 1 litre.

c. NaCl solution—add 1.5 g of AR grade NaCl to 100 ml of water.

d. Standardisation of $Hg(NO_3)_2$ solution—add 1 ml of diphenylcarbazone indicator to 25 ml of the NaCl solution and titrate against $Hg(NO_3)_2$ solution till permanent blue–purple colour appears.

e. Indicator solutions—$Fe(NO_3)_3$ is prepared by dissolving 5 g of $Fe(NO_3)_3$ in 100 ml of distilled water. Na-thiocyanate solution is prepared by dissolving 0.008 g of Na-thiocyanate in 100 ml of distilled water.

Procedure

i. Take 25 ml of aliquot in a 50 ml graduated flask.

ii. Add 0.5 ml 1 M HNO_3 and 20 ml Na-tetraphylborate solution.

iii. Pour the mixture into a 150 ml flask with stopper.

iv. Shake well for 5 minutes and filter through Whatman No. 40 filter paper.

v. Transfer 25 ml of the filtrate into a 250 ml conical flask. Add 75 ml distilled water, 1 ml $Fe(NO_3)_3$ solution and 1 ml Na-thiocyanate solution.

vi. Titrate against $Hg(NO_3)_2$ till a colourless end point is reached from which the quantity of Na-tetraphylborate that remained unconsumed is found out.

vii. Determine the amount of K_2O present in stock solution from the amount of Na-tetraphylborate that took part in chemical reaction.

The estimation of Na_2O and K_2O by the above methods does not give satisfactory results always. Hence, the percentages of Na_2O and K_2O are determined by flame photometry.

3.8.4. Flame Photometry

Principle: Flame transfers solid or liquid into vapour states and decomposes them to simpler molecules or atoms. The vapours of neutral metal atoms or molecules containing atoms are excited by thermal energy of the flame. From the excited levels of atoms, the electrons tend to return to the ground state by emission of radiation. A particular element would give a characteristic spectrum of its own. The electrons return to the ground state in stages leading to generation of several spectral lines. Though energy level diagrams can be constructed for all atoms, those for mono- and diatomic molecules like Na, K, Ca, etc. are relatively simpler. In such cases, the most outstanding line is due to transition of electrons between the lowest excited state and ground state. For alkali metals Li, Na, K and Ca such transitions occur at 671 nm, 589 nm, 766.5 nm and 423 nm, respectively. It is possible to regulate the energy level transitions by controlling the temperature of the flame. However, the intensities of spectral lines are dependent on the concentration of the element.

A block diagram of flame photometer is shown in Fig. 3.1. The essential parts of this instrument are pressure regulator and flow meter, atomizer, burner, photosensitive detector and recording output of the detector. The pressure regulator and flow meter are used for proper adjustment of pressure and flow of gasses. The atomizer is used to introduce liquid sample into the flame at a stable rate. The capillary-aspirator type atomizer is advantageous as it atomizes solution at a slow rate (4 ml per minute). It draws the solution up through the capillary intake tube from a small beaker. When the beaker is withdrawn, the capillary tube empties instantaneously. The outside of

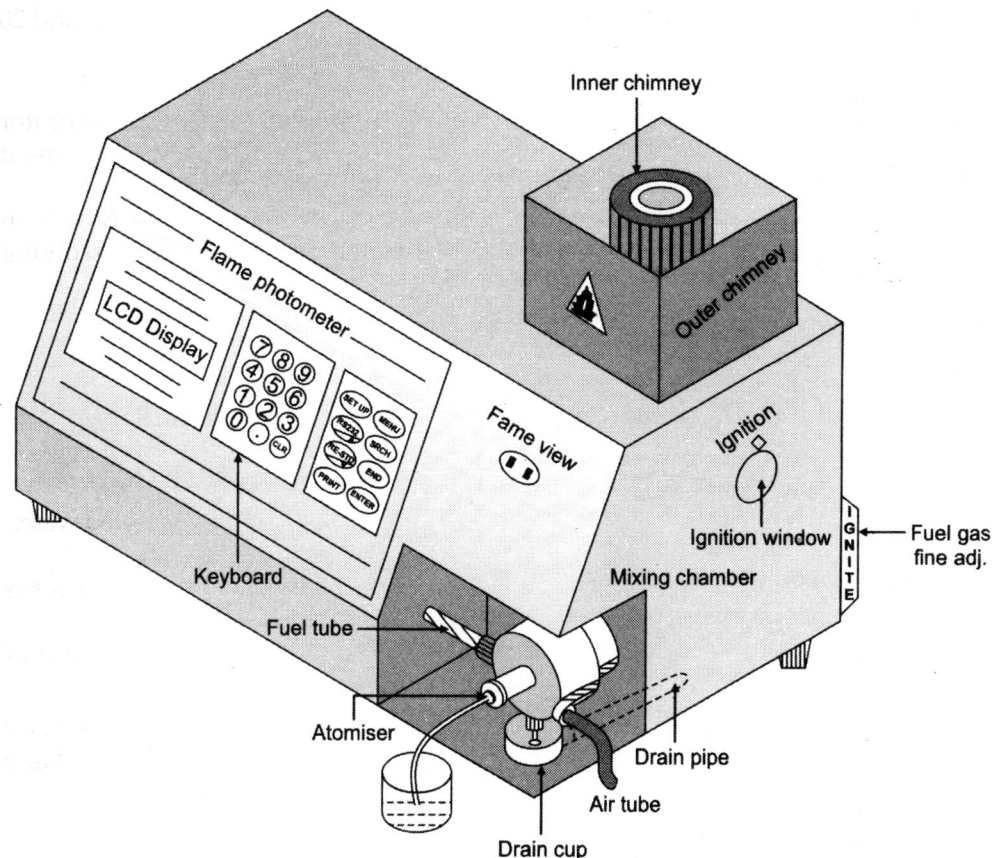

Fig. 3.1: Block diagram of the Systronics flame photometer 128

the tube can be wiped clean with a piece of tissue paper. The capillary tube is flushed clean of the previous solution as a new solution is fed. Flame photometer 128 with compressor 126 made by Systronics is one of the standard instrument for estimation of Na_2O and K_2O.

3.8.5. Estimation of Na_2O and K_2O by Flame Photometry

a. Preparation of standard solution:
   i. For 1000 ppm Na—dissolve 2.542 g NaCl in 1 litre distilled water.
   ii. For 1000 ppm Na_2O—dissolve 1.886 g NaCl in 1 litre distilled water.
   iii. For 1000 ppm K—dissolve 1.907 g KCl or 2.5869 g KNO_3 in 1 litre distilled water.
   iv. For 100 ppm K_2O—dissolve 1.583 g KCl in 1 litre distilled water.
   v. Dilute the solution to make 100 ppm, 80 ppm, 60 ppm, 40 ppm and 20 ppm standard solutions.
b. Operating procedure:
   i. Ensure that air-tube, gas tube and drain tube are properly connected (Fig. 3.1).

ii. Switch on the compressor. Ensure that the output pressure is 0.5 kg/cm^2. If the pressure is more or less than 0.5 kg/cm^2, make necessary adjustments according to the instruction manual.

iii. Dip the atomizer capillary tube in distilled water. Water drops will fall in the drainage cup (Fig. 3.1).

iv. Open the fuel gas fine adjustment valve approximately half turn in the direction indicated by arrow.

v. Switch on the fuel supply from the LPG cylinder (fuel source) and immediately ignite the flame through the ignition window.

vi. Watch the flame through the flame view window. Make fine adjustment of flow of fuel with the help of the gas control valve to get a stable flame having well defined blue cone.

vii. For selection of elements (Na or K) press the assigned numerical key followed by 'ENTER' key in the 'Edit' menu.

viii. Subsequently the display will ask for 'STANDARD SOLUTIONS' and 'DILUTION FACTOR' to be used. After these values are entered, the display will go to 'SET UP' menu.

ix. Set up numbers between 1–20 for 'mono chromatic' and 21–40 for 'bi-chromatic' modes.

x. There are two operating modes: 1—for low concentration and 2—for high concentration. Since the rocks and minerals with Na and/or K as the major element contain appreciably higher amount of Na and/or K, the instrument is to be set up for the 'high concentration' mode.

xi. Calibrate the instrument for 20–100 ppm by following the instructions given in the instrumentation manual.

xii. After calibration is over, the instrument is ready for sample analysis. Take the solution for which the estimation is made in sample beaker and dip the capillary tube of the atomiser. LCD display will indicate the concentration of element or oxide in solution, which is to be calculated to get the weight percent in original rock or mineral sample. In case of any difficulty, refer to the instruction manual provided by the manufacturer.

3.9. ESTIMATION OF TiO$_2$ BY FERRIC AMMONIUM SULPHATE TITRATION METHOD

i. Take 50 ml aliquot in a 500 ml conical flask; add 25 ml conc. HCl and a strip of 2 g aluminium foil.

ii. Boil the mixture on a burner in the presence of CO_2 for at least 30 minutes till the entire aluminium foil is digested. A slow stream of CO_2 is to be maintained.

iii. Cool the flask in cold water (preferably ice cold water) with a slow current of CO_2.

iv. Add 5 ml ammonium thiocyanate solution to the content of the flask, shake well and titrate quickly with standard 1 N ferric ammonium sulphate solution till a faint red colour appears.

1 ml 1 N ferric ammonium sulphate = 0.08 g TiO$_2$

3.10. ESTIMATION OF MnO BY PERMANGANATE METHOD

i. Take 25 ml aliquot in a 500 ml conical flask and boil to remove chlorine, if present.

ii. Add 125 ml of demineralised water and 5–6 g of ZnO to the flask and shake well.

iii. Heat the mixture on a burner (boiling is to be avoided).
iv. Titrate the content of the flask against standard $KMnO_4$ solution to a permanent pink colour.

$$1 \text{ ml } 1 \text{ N } KMnO_4 = 0.01648 \text{ g of Mn} = 0.02128 \text{ g of MnO}$$

3.11. ESTIMATION OF P_2O_5 BY WILSON METHOD

i. Dissolve 0.05–0.2 g of sample in 25 ml conc. HCl in a 500 ml conical flask by heating slowly on a burner and add few drops of HNO_3.
ii. Cool the content and neutralise the acid with 10% NaOH solution. Litmus paper can be used as an indicator.
iii. Redissolve the precipitate formed in dil. HCl. Add about 5 ml dil. HCl in excess.
iv. Dilute the solution to 150 ml with demineralised water; add a pinch of citric acid to the solution and boil on a burner.
v. Add 30 ml of citromolybdate solution to the content and boil.
vi. Add 25 ml quinoline hydrochloride solution slowly with constant stirring and boil the solution for 5–10 minutes.
vii. Cool the content by cold-water bath overnight. Precipitation will take place.
viii. Collect the precipitate on a pad of filter paper pulp under suction.
ix. Wash the conical flask thoroughly with cold water and transfer the water to the precipitate collected on the pad.
x. Wash the precipitate 8–10 times with cold water till it is free from acid.
xi. Transfer the precipitate along with filter paper pad to the original flask.
xii. Add 50 ml of demineralised water and 0.1N NaOH (4 g NaOH in 1 liter distilled water) solution to the flask to dissolve the yellow precipitate of quinoline phosphomolybdate completely. Add 2–3 drops of phenolphthalein indicator.
xiii. Titrate the content of the flask against 0.1N HCl till the solution becomes colourless.

$$1 \text{ ml } 1\text{N HCl} = 0.001193 \text{ g of P} = 0.002733 \text{ g of } P_2O_5$$

3.12. ESTIMATION OF WATER

Water is available in form of water of crystallization within the crystal lattice and in adsorbed (hygroscopic) form.

3.12.1. Estimation of Water of Crystallization (H_2O+) by Penfield's Tube

i. Dry the sample at 110°C for 1 hour.
ii. Take a dry Penfield's tube (Fig. 3.2a) and determine its weight (W_1).
iii. Pour about 1 g of the sample into the lower bulb through a capillary funnel (Fig. 3.2b).
iv. Let the weigh of the Penfield's tube with sample is W_2. The exact weight of the sample is (W_2-W_1).
v. Hold the tube horizontally; heat the first bulb on flame to red hot while wrapping the second bulb by a moist filter paper or cloth.
vi. Pull out and cut off the heated bulb along with the residue from the rest of the tube with flame and seal the cut end of the tube.
vii. Cool the tube, clean the external portion and weigh (W_3).

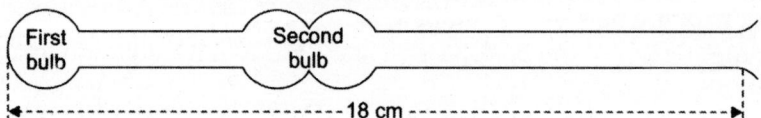

Fig. 3.2a: Penfield's tube

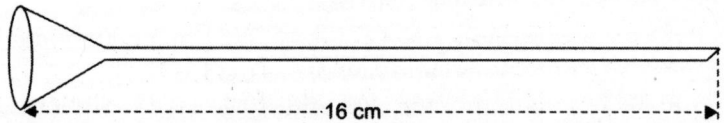

Fig. 3.2b: Capillary funnel

viii. Place the tube inside an electric oven at 100°C for 15 minutes.
  ix. Remove the tube from the oven, cool in a desiccator and weigh (W_4).
   x. The difference in weight ($W_3 - W_4$) is the weight of water.

$$H_2O+ \ (\%) = \frac{W_3 - W_4}{W_2 - W_1} \times 100$$

3.12.2. Estimation of Adsorbed Water (H₂O–)

   i. Take about 1 g of sample in a 30 ml porcelain crucible and weigh (W_1).
  ii. Heat the crucible in an oven at 105 to 110°C for 1 hour.
 iii. Transfer the crucible into a desiccator; allow to cool for 30 minutes and weigh (W_2) again.
  iv. The difference in weight ($W_1 - W_2$) is the weight of water.

$$H_2O-(\%) = \frac{W_1 - W_2}{W_1} \times 100$$

3.13. Estimation of CO₂

   i. Set up the instruments as shown in Fig. 3.3.
  ii. Take about 0.5 (W_1) g of the sample in the flask and cover it with water.
 iii. Close the stopper cock in the separating funnel and insert the weighed (W_2) absorption bulb in the system and put the guard tube in position.
  iv. Allow air free from CO₂ to pass through the system.
   v. Half fill the separating funnel with 1:1 HCl.
  vi. Open the stopcock of the dropping funnel so that acid can go into the flask slowly.
 vii. Start a flow of water in the condenser when effervescence diminishes and heat the flask slowly so as to secure steady but quiet bubbling.
viii. Remove the flame when it is seen that CO₂ has been boiled out of the solution. Increase the current of air free from CO₂ that will sweep out all CO₂.
  ix. Close the inlet and outlet tubes and disconnect the weighed absorption bulb.
   x. Cool the absorption bulb in a desiccator and weigh (W_3).

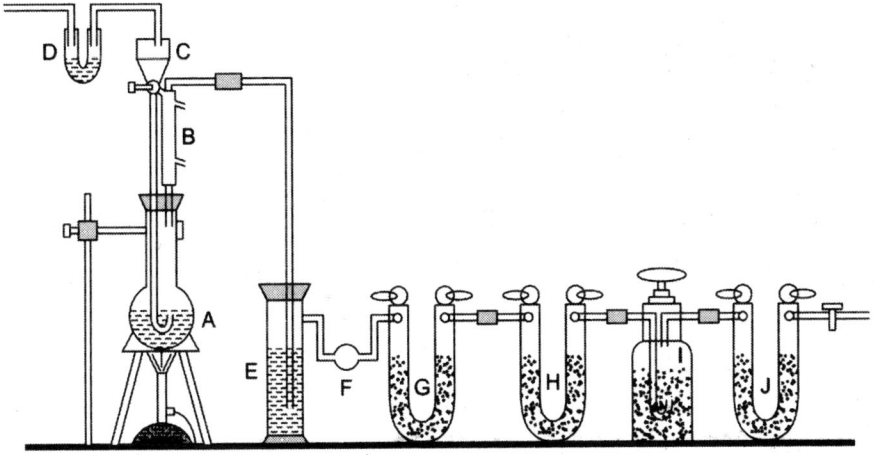

index

A — 250 ml flask
B — Condenser
C — Dropping funnel
D — Soda lime guard tube
E — Bubble

F — Absorption bulb
G— U-tube with CuSO4
H— U-tube with Mg-perchlorate
I — Valve tube with ascarite and anhydrone
J — Guard tube

Fig. 3.3: CO_2 apparatus

$$CO_2\,(\%) = \frac{W_3 - W_2}{W_1} \times 100$$

where,

W_1 = weight of the sample
W_2 = weight of the bulb before the test
W_3 = weight of the bulb after the test

3.14. ESTIMATION OF S

i. Take 0.1 to 0.2 g of sample in a nickel crucible and thoroughly mix about 2 g of sodium peroxide with a dry glass rod.

ii. Cover the mixture with a thin layer of the peroxide and heat gently over a low flame for 10 to 15 minutes at a temperature just sufficient to produce complete fusion.

iii. Keep the melt over the flame for 5 minutes and give a swirling motion to the crucible 2 or 3 times with the help of tongs.

iv. Remove the crucible from heat and allow it to cool in air slowly. Cooling in water may cause an explosion.

v. Place the crucible in a 500 ml beaker, pour 100 ml water to it and cover quickly with a watch glass to prevent loss by spattering.

vi. Boil the beaker for a few minutes with constant stirring to destroy excess of peroxide.

vii. If the supernatant liquid is greenish, it indicates the presence of manganese. In such case add 2 to 5 ml methyl alcohol and boil to precipitate manganese as MnO_2.

viii. Filter the hot solution through Whatman No. 40 filter paper into a 500 ml beaker using a funnel of large size so that the layer of precipitate on the filter paper will not be too thick.

ix. Wash the residue 6 to 8 times with hot water. The total filtrate and washings should be about 250 to 300 ml.

x. Add a few drops of methyl red and about 3 ml HCl.

xi. Heat the beaker till the solution boils. Add 10–20 ml of 10% $BaCl_2$ solution slowly with constant stirring. Keep it hot for a few minutes.

xii. Remove the beaker from the flame and keep overnight to allow the precipitate ($BaSO_4$) to settle.

xiii. Filter the content through Whatman No. 42 filter paper and wash the precipitate 8 to 10 times with hot water.

xiv. Transfer the residue along with the filter paper to a weighed platinum crucible and ignite the residue carefully at a temperature of 1000°C.

xv. Take the residue out of heat; cool in a desiccator and weigh again.

xvi. Calculate the percentage of S in the sample from the weight of the $BaSO_4$ by the following formula.

$$S\,(\%) = \frac{13.74 \ \times Wt.\ of\ BaSO_4}{Wt.\ of\ sample}$$

3.15. ESTIMATION OF F

Reagents

Following reagents are necessary for the estimation of fluorine. These should be freshly prepared.

a. **Acid mixture:** Conc. H_2SO_4 of specific gravity 1.84 + Conc. H_3PO_4 (85%) in 3:1.

b. **Standard thorium nitrate solution:** Dissolve 13.086 g of $Th(NO_3)_4 \cdot 4H_2O$ in 1 litre distilled water and standardize against standard fluoride solution.

c. **Monochloroacetic acid:** Dissolve 18.9 g of monochloroacetic acid in 200 ml distilled water. Neutralise 100 ml with NaOH and add to the rest. Dilute to 500 ml with distilled water. The pH should be within 2.9 to 3.1.

d. **Sodium salt of alizarine sulphonate:** 1:1 aqueous solution.

The procedure of fluorine estimation is given below.

i. Take 1 g of sample and 1–2 g of Na_2CO_3 in a platinum crucible and fuse.

ii. Take the fused mass in a beaker, add 50–100 ml of water and filter the mixture.

iii. Transfer the filtrate to the Kjeldahl flask of fluorine distillation apparatus (Fig. 3.4). Add 75 ml acid mixture and 1 g of sodium silicate or quartzite. Heat the mixture on an asbestos sheet with a small hole at the center.

iv. When the temperature reaches 140°C, flush steam to collect the distillate in 25 ml of 0.1 N NaOH (4 gram NaOH in 1 liter distilled water) solution. The distillate should be about 300–350 ml.

v. Take 20 ml of aliquot and titrate against 0.025 M $Th(NO_3)_4$ solution using 10 ml monochloroacetic acid as a buffer and few drops of sodium salt of alizarine sulphonate as an indicator.

vi. Determine the percentage of fluorine from the volume of $Th(NO_3)_4$ used.

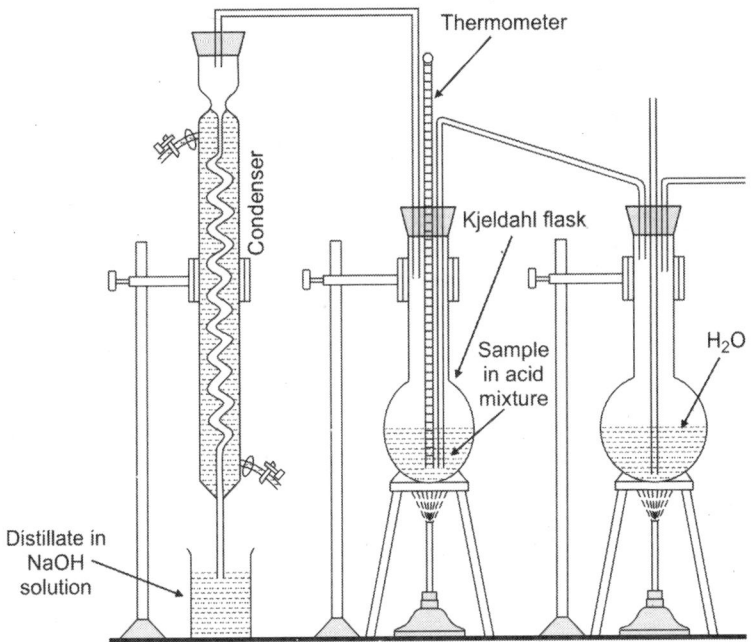

Fig. 3.4: Fluorine distillation apparatus

3.16. ESTIMATION OF Cl BY VOLHARD'S METHOD

The mineral powder solution is treated with excess $AgNO_3$ solution. The amount of residual $AgNO_3$ is determined by titration with standard thiocyanate solution. The procedure is given below.

   i. Pipette out 25 ml of aliquot into a 250 ml conical flask.
   ii. Add 5 ml of 6 M HNO_3 and about 30 ml of 0.1 M $AgNO_3$ (17 g $AgNO_3$ in 1 liter distilled water) solution (sufficient to give 2.5 ml in excess after reaction)
   iii. Boil the mixture. White AgCl will be precipitated.
   iv. Cool the flask.
   v. Filter the content of the flask through a quantitative filter paper or porous porcelain or sintered glass crucible.
   vi. Add 1 ml of Fe^{+3} indicator to the filtrate and titrate the residual $AgNO_3$ against standard 0.1 M thiocyanate till colour changes.
   vii. Calculate the volume of standard 0.1 M $AgNO_3$ that has reacted with the chloride solution.

3.17. ESTIMATION OF LOI

   i. Take about 1 g (W_1) of dry sample in a silica or platinum crucible.
   ii. Weigh the crucible with sample (W_2).
   iii. Heat the crucible in a muffle furnace for 30 minutes at temperature of 300°C.
   iv. Take the crucible out of furnace and cool in a desiccator.
   v. Weigh the crucible with dry sample (W_3).

$$\text{Loss of ignition (LOI) (\%)} = \frac{W_2 - W_3}{W_2 - W_1} \times 100$$

Quantitative Analysis of Ores and Minerals

Quantitative estimation of metal contents of ores and valuable constituents of nonmetallic economic materials are necessary for proper assessment of mineral wealth. The procedures of estimation of Sb, As, Co, Cr, Fe, Mn, Mo, Ni, Cu, Pb, Zn, Sn, Ti, V, W and Zr in respective ores as well as estimation of $BaSO_4$ in barite, CaF_2 in fluorite, CaO and P_2O_5 in apatite and $CaSO_4.2H_2O$ in gypsum are described in this section.

4.1. ESTIMATION OF ANTIMONY IN ANTIMONY ORE

Stibnite (Sb_2S_3) is the ore and chief commercial source of antimony even though minerals like tetrahedrite ($Cu_{12}Sb_4S_{13}$), boulangerite ($Pb_5Sb_4S_{11}$) and bournonite (PbCuSbS$_3$) contain appreciable amounts of antimony. The procedure for estimation of antimony in antimony ores is as follows:

i. Take about 0.5 to 1 g of finely powdered ore in a 50 ml pyrex or corning conical flask.

ii. Add 3–5 g of K_2SO_4, 15 ml conc. H_2SO_4 and a piece of paper to the flask and heat gradually initially at low temperature and finally with a full flame.

iii. Continue heating till a clear solution is obtained.

iv. Cool the flask by rotating gently.

v. Dissolve the melt in 50 ml 1:1 HCl by gentle warming.

vi. Transfer the contents of the flask to a 500 ml beaker; rinse the original flask with 25 ml conc. HCl.

vii. Pass a current of H_2S through the solution to precipitate arsenic sulphide.

viii. Filter the content through a double filter paper moistened with 2:1 HCl. Wash the precipitate 5–6 times with 2:1 HCl. Antimony passes into the filtrate together with the other elements present in the ore.

ix. Dilute the filtrate with three times of its volume of warm water and then saturate with H_2S gas. Antimony sulphide together with the other elements of the hydrogen sulphide group will be precipitated.

x. Filter the mixture through Whatman No. 40 filter paper and wash the residue with dilute H_2SO_4 saturated with H_2S. Reject the filtrate.

xi. Treat the precipitate on the filter paper with 10 to 20 ml mixture of AR grade Na_2S and NaOH (70 g of Na_2S with 40% of NaOH diluted to 1000 ml). Antimony sulphide will be separated from Cu-, Pb-, Cd- and Bi-sulphides.

xii. Add 2 g of Na_2SO_4 and 10 ml conc. H_2SO_4 to the filtrate and heat the washings containing antimony until sulphur is destroyed and most of the free acid is flushed out.

xiii. Add 30–35 ml HCl to the filtrate and dilute to 200 ml with water.

xiv. Titrate the filtrate with 0.1 N $KMnO_4$ solution (3.3 gram of $KMnO_4$ in 1 liter distilled water) until pink colour persists for 10 to 30 seconds.

xv. Amount of antimony (Sb) is calculated by following equivalence.

$$1 \text{ ml of } 1 \text{ N } KMnO_4 = 0.06087 \text{ g of Sb}$$

4.2. ESTIMATION OF ARSENIC IN ARSENIC ORE

The principal minerals of arsenic are arsenopyrite (FeAsS), realgar (AsS) and orpiment (As_2S_3). Many arsenides and sulphoarsenides of lead, copper, gold and tin also contain arsenic. The procedure for estimation of arsenic in arsenic ores is as follows:

i. Take about 0.5 to 5 g finely powdered sample in a beaker.

ii. Add 15 ml of fuming HNO_3 and digest the ore by heating the mixture on a hot plate at low temperature. Add 10 ml 1:1 H_2SO_4.

iii. Cool the solution, dilute with water and transfer the entire solution into a distillation flask.

iv. Add 1 g of hydrazine sulphate and 0.5 g of cuprous chloride to the flask to reduce the arsenic to trivalent state.

v. Add 70 ml conc. HCl to the flask and set up the distillation apparatus.

vi. Increase the temperature slowly to 100–110°C and collect about 100 ml distillate in the collecting beaker.

vii. Transfer the distillate to a 500 ml beaker, cool, make slightly alkaline with 20% NaOH and then acidify slightly with HCl. Ascertain the alkaline and acidic conditions by putting a piece of litmus paper in the beaker.

viii. Add 3–4 g of $NaHCO_3$ and a little starch solution to the beaker.

ix. Titrate with standard iodine solution till permanent blue tinge appears.

$$1 \text{ ml } 1 \text{ N iodine} = 0.03746 \text{ g of As} = 0.04946 \text{ g of } As_2O_3$$

4.3. ESTIMATION OF COBALT IN COBALT ORE

The chief ore of cobalt is cobaltite [(Co, Fe)AsS]. Other common minerals are linnaeite (Co_3S_4), carroltite ($CuCo_2S_4$), etc. Cobalt is estimated from cobalt ores by the gravimetric method, which is given below.

i. Take 0.5 to 1 g (W_1) of finely powdered sample in a 250 ml beaker.

ii. Add 10–25 ml conc. HNO_3, 10–15 ml conc. HCl, and 10 ml 1:1 H_2SO_4 and a few drops of HF. Boil the mixture gently over a hot plate and finely vigorously to evaporate strong fumes of SO_3.

iii. Cool the solution, dilute with water and boil again.

iv. Acidify the solution with 5 to 10% H_2SO_4 and pass a strong stream of H_2S through the solution.

v. Filter the solution through Whatman No. 40 filter paper and reject the residue.

vi. Boil the filtrate for 15 minutes to remove H_2S.

vii. Add 25 ml of H_2O_2 to oxidize the iron and boil the solution to remove excess peroxide.

viii. Neutralise the solution by Na_2CO_3 solution and boil to remove all CO_2.

ix. Add zinc oxide suspension to the solution till the precipitate becomes coffee coloured.

x. Boil the mixture, cool and filter through Whatman No. 40 filter paper with some pulp.

xi. Add 10 ml conc. HCl to clear the solution.

xii. Dilute the solution to 200 ml, heat to about 60°C and add α nitroso β napthol slowly with constant stirring.

xiii. Cool the solution for several hours to bring it to room temperature.

xiv. Filter the content through Whatman No. 42 filter paper, using pulp, and wash the precipitate several times with warm dilute HCl.

xv. Transfer the paper and precipitate to a deep type weighed 30 ml porcelain crucible.

xvi. Ignite the paper and precipitate gently at first and finally to constant weight at 750 to 850°C.

xvii. Cool the resultant solid in a desiccator and weigh (W_2).

$$Co\ (\%) = \frac{73.42 \times W_2}{W_1}$$

4.4. ESTIMATION OF CHROMIUM IN CHROMITE

Chromite is the most important ore of chromium. Chemically, it is represented by $FeCr_2O_4$, but Mg and Al are found in the lattice replacing Fe and Cr, respectively. The procedure of estimation of Cr_2O_3 is given below.

i. Take 0.2 g of finely grinded chromite powder in a nickel crucible.

ii. Add about 1 g of sodium peroxide to the mineral powder and mix both thoroughly with a glass rod.

iii. Heat the crucible gently over a low flame till the mass melts.

iv. Cool the crucible and place it in a 500 ml beaker containing a little water. Cover the beaker by a watch glass.

v. Add nearly 100 ml of water to the beaker and after completion of the violent reaction remove the crucible with a glass rod and wash thoroughly. Pour the washed water into the same beaker.

vi. Boil the solution, cool to room temperature and filter through Whatman No. 40 filter paper.

vii. Wash the residue with hot water till it is free from chromate ions.

viii. As there are chances of some chromium being held up by the residue, dissolve the residue in the filter paper by minimum amount of HCl.

ix. Dilute the solution and add sodium peroxide slowly with constant stirring till a permanent precipitate is obtained.

x. Boil the content, cool and filter through Whatman No. 40 filter paper.

xi. Wash the residue 5–6 times with hot water.

xii. Mix the filtrate with the filtrate obtained in step (vi) above.

xiii. Boil the filtrate for half an hour to decompose hydrogen peroxide.

xiv. Cool the filtrate, acidify with 1:1 H_2SO_4 and add measured amount of ferrous ammonium sulphate solution till the solution becomes green.

xv. Add 5 ml of orthophosphoric acid to the solution and titrate the excess of ferrous ammonium sulphate with 0.1 N potassium dichromate solution using barium diphenylamine sulphonate as indicator until the colour changes to violet.

xvi. The actual volume of the ferrous ammonium sulphate solution, which was oxidised by the dichromate originating from chromite, is determined by following factor.

$$1 \text{ ml } 1 \text{ N } K_2Cr_2O_7 = 0.02534 \text{ g of } Cr_2O_3 = 0.01734 \text{ g of Cr}$$

4.5. ESTIMATION OF IRON IN IRON ORE

Hematite (Fe_2O_3) and magnetite (Fe_3O_4) are common iron ores. The procedure of estimation of Fe in hematite is given below. Since Fe occurs both in Fe^{+3} and Fe^{+2} states, the Fe^{+3} is converted to Fe^{+2} before final estimation.

i. Take about 0.5 g of fine powdered sample in a conical flask; add 100 ml of dilute HCl (1:1) and warm gently. Continue heating till no colour particle is left out.

ii. Cool the flask and filter the liquid through Whatman No. 40 filter paper into a 250 ml flask.

iii. Wash the left out residue with dilute HCl and pass the washings into the volumetric flask.

iv. Make the solution 250 ml by adding distilled water and shake the flask well.

v. Pipette out 50 ml of the solution to a conical flask, heat to boiling.

vi. Add a few drops of $SnCl_2$ to the hot solution with constant stirring till the yellow colour of the solution nearly disappears.

vii. Add a few drops of hot $SnCl_2$ to the solution in excess till the solution acquires a faint green colour free from any tinge of yellow colour. (If the colour changes to brown or white due to excess addition of $SnCl_2$, reject the solution and repeate the above procedures.)

viii. Cool the solution rapidly under tap water covering it with a watch glass.

ix. Remove the excess $SnCl_2$ by adding 10 ml of $HgCl_2$ rapidly in one portion of the flask.

x. A slight silky white precipitate will be formed. Formation of heavy, gray or black precipitate indicates addition of excess $SnCl_2$. In such case, the experiment is to be repeated again.

xi. Allow the solution to stand for 5 minutes and then transfer to a 1 litre beaker.

xii. Add 400 ml distilled water and 25 ml of Zimmermann-Rainhandt solution (Z-R solution).

xiii. Titrate 50 ml of the solution against 0.1 N $KMnO_4$ solution (3.3 gram of $KMnO_4$ in 1 liter distilled water) with constant stirring till the entire solution assumes pink colour, which is stable for at least 15 seconds.

xiv. Titration may be repeated twice.

Calculation

Say 10 ml of $KMnO_4$ is necessary in titration.

Strength of the $KMnO_4$ solution = 0.1 N

1 ml of 1N solution of $KMnO_4$ = 0.05584 g of iron

1 ml of 0.1N solution of $KMnO_4$ = 0.005584 g of iron

10 ml of 0.1N solution of $KMnO_4 = (0.005584 \times 10) = 0.05584$ g of iron

So 50 ml of solution contains 0.05584 g of iron.

250 ml of solution contains $0.05584 \times 5 = 0.2792$ g of iron

0.5 g of sample contains 0.2792 g of iron.

100 g of sample contains $= (0.2792 \div 0.5) \times 100 = 55.84$ g of iron

Conclusion: The iron content of the given iron ore powder is 55.84%.

4.6. ESTIMATION OF FeS$_2$ IN PYRITE

i. Digest 1 g of powdered pyrite sample in 25 ml conc. HCl for about fifteen minutes in a beaker.

ii. Filter the content through Whatman No. 40 filter paper and wash the residue with hot water.

iii. Ignite the residue in a silica crucible and dissolve in conc. HCl.

iv. Add $SnCl_2$ to the hot solution.

v. Cool the solution and add 10 ml mercuric chloride.

vi. Make the volume nearly 200 ml with distilled water.

vii. Add 25 to 30 ml of sulphuric phosphoric acid mixture and titrate the solution with 0.1N potassium dichromate solution using few drops of barium diphenylamine sulphonate as an indicator.

viii. Determine the percentage of iron and calculate S and FeS$_2$ by following formulae.

1 ml 1N $K_2Cr_2O_7 = 0.05584$ g Fe

S % = $1.1484 \times$ Fe% and FeS$_2$ % = $2.1484 \times$ Fe%

4.7. ESTIMATION OF MANGANESE IN MANGANESE ORE

Chief ore minerals of manganese are pyrolusite and psilomelane. The procedure of estimation of Mn in pyrolusite (MnO_2) is as follows:

i. Take exactly 1.7 g of AR sodium oxalate in a 250 ml volumetric flask and make up to the mark by adding distilled water; shake well.

ii. Take 25 ml of the solution in a 400 ml conical flask and add 150 ml of 2N H_2SO_4.

iii. Titrate against 0.1N standard $KMnO_4$ solution till pink colour appears throughout the solution.

iv. Allow the solution to remain as such till it becomes colourless.

v. Warm the solution up to 50–60°C and titrate against standard $KMnO_4$ solution with constant stirring till a permanent faint pink colour appears that persists for at least 30 seconds. (Say 25 ml of Na-oxalate solution consumes 26 ml of $KMnO_4$)

vi. Take exactly 0.2 g of finely powdered dry pyrolusite in a conical flask.

vii. Add 50 ml of standard 0.1N Na-oxalate solution and 50 ml of 4N H_2SO_4.

viii. Put a short funnel on the mouth of the flask.

ix. Boil the mixture gently until it is completely free from black particles.

x. Cool the solution and titrate excess oxalate with standard $KMnO_4$ solution. (Say excess amount of Na-oxalate solution consumes 12 ml of $KMnO_4$)

xi. Calculate the amount of Na-oxalate consumed in the reaction and determine the percentage of MnO_2 in ore.

Calculation

Volume of Na-oxalate solution (V_1) = 25 ml

Let the volume of $KMnO_4$ solution consumed in titration against Na-oxalate solution (V_2) = 26 ml

Strength of $KMnO_4$ solution (S_2) = 0.1N

Strength of Na-oxalate solution = $(V_2 \times S_2) \div V_1 = (26 \times 0.1) \div 25 = 0.104$ N

Let the volume of $KMnO_4$ solution consumed in titration against ore solution = 12 ml

Volume of Na-oxalate solution consuming this 12 ml of $KMnO_4$ = $(12 \times 0.1) \div 0.104 = 11.54$ ml

So 11.54 ml of Na-oxalate solution remained unconsumed by MnO_2.

Na-oxalate consumed by MnO_2 = 50 – 11.54 = 38.46 ml

1 ml of 1N Na-oxalate solution = 0.04346 g of MnO_2.

38.46 ml (0.104 N) of Na-oxalate solution = $0.04346 \times 38.46 \times 0.104 = 0.173833$ g of MnO_2

0.2 g of pyrolusite contains 0.173833 g of MnO_2.

100 g of pyrolusite contains = $(0.173833 \times 100) \div 0.2 = 86.9165$ g of MnO_2

$(54.9 + 32)$ = 86.9 g of MnO_2 contains 54.9 g of Mn

86.9165 g of MnO_2 contains = $(54.9 \times 86.9165) \div 86.9 = 54.91$ g of Mn

Conclusion

The Mn content in pyrolusite is 54.91%.

4.8. ESTIMATION OF MOLYBDENUM IN MOLYBDENUM ORE

Important ore minerals of molybdenum are molybdenite (MoS_2), molybdite (MoO_3), wulfenite $(PbMoO_4)$, powellite {$Ca(Mo,W)O_4$}, etc. The procedure of estimation of Mo is given below.

  i. Take 0.2 to 0.5 g of finely pulverized sample and approximately 5 times its weight of sodium peroxide in a nickel crucible and mix thoroughly by a dry glass rod.

 ii. Heat the mixture over a low flame to obtain a clear melt.

iii. Swirl the crucible from time to time to include all the black particles.

 iv. Cool the crucible, put in a 500 ml beaker and pour nearly 200 ml distilled water into the beaker; boil the contents.

  v. Remove the crucible from burner and wash well with water. Pour the washings into the beaker.

 vi. Filter the content of the beaker through Whatman No. 40 filter paper and discard the precipitate.

vii. Neutralise the filtrate with 1:1 H_2SO_4. Add 20 ml H_2SO_4 in excess.

viii. Activate the Jone's reductor (Fig. 4.1) by passing about 400 ml of 4% H_2SO_4 through it. Discard the washings.

 ix. Place a 500 ml conical flask containing 25 ml of 10% ferric ammonium sulphate solution and 5 ml orthophosphoric acid.

  x. Pass CO_2 through the flask having an inlet and outlet for the CO_2 gas.

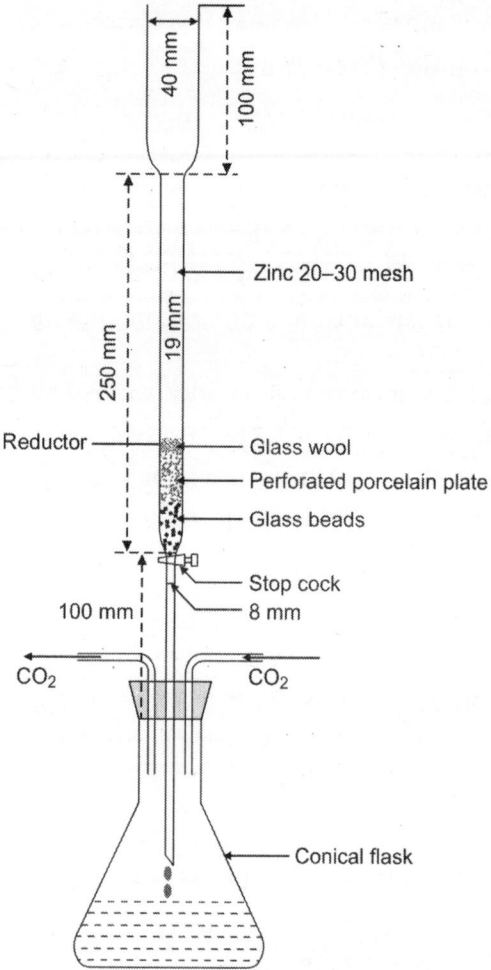

Fig. 4.1: Jone's reductor

xi. Pass the solution through the reductor at a slow rate such that 2 or 3 drops will fall per second into the conical flask.

xii. Wash the reductor with 4% H_2SO_4 several times at the end and collect all the washing in the conical flask. Maintain the stream of CO_2 throughout the experiment.

xiii. Remove the conical flask, add 2 ml phosphoric acid and titrate the solution against standard $KMnO_4$ solution to a pink end point, which should persist for at least 5 seconds.

$$1 \text{ ml } 1 \text{ N } KMnO_4 = 0.032 \text{ g of Mo}$$

4.9. ESTIMATION OF NICKEL IN NICKEL ORE

The chief nickel bearing minerals are niccolite (NiAs), pentlandite (Ni,Fe)S, garnierite $(Ni,Mg)SiO_4.nH_2O$, millerite (NiS), chloanthite $(NiAs_2)$, etc. Nickel is also found associated with copper ore deposits and laterites.

4.9.1. Estimation of Nickel from Sulphide Ores

i. Take 0.5 to 1 g (W_1) of the finely powdered sulphide sample in an iron crucible and add approximately equal amount of sodium peroxide.

ii. Dissolve the mixture in HCl and pass H_2S through the solution till precipitation takes place.

iii. Filter the mixture and wash the residue with water saturated with H_2S.

iv. Boil the filtrate vigorously to remove H_2S; add bromine water and boil the solution to oxidize ferrous iron to the ferric state.

v. Neutralise the solution with NH_4OH and pyridine to precipitate iron.

vi. Filter the mixture through Whatman No. 40 filter paper.

vii. Wash the residue 5–6 times with hot ammonical water.

viii. Acidify the filtrate with HCl, add 10 ml of 10% dimethyl glyoxime and heat on a burner.

ix. Keep the solution over a boiling water bath for 2 hours and then allow the precipitate to cool overnight.

x. Filter the mixture through a weighed sintered glass crucible, wash with cold water and then dry the precipitate at 150°C.

xi. Cool the precipitate in a desiccator and find out the weight of the precipitate (W_2).

$$Ni\ (\%) = \frac{20.31 \times W_2}{W_1}$$

4.9.2. Estimation of Nickel from Oxidised Ores

i. Take 0.5 to 1 g (W_1) of sample in a beaker, add 15 to 20 ml conc. HCl and 5–10 ml HNO_3, dry and dehydrate the sample on a burner.

ii. Add HCl, boil and filter through Whatman No. 40 filter paper.

iii. Ignite the residue in a platinum crucible and add HF.

iv. Fuse the residue with potassium pyrosulphate and dissolve in HCl and add to filtrate.

v. If copper is present, remove it with H_2S.

vi. Add bromine water and boil to oxidise iron.

vii. Add 2 g of tartaric acid and NH_4OH to prevent the precipitation of the hydroxides of iron, aluminium and chromium.

viii. Then slowly make the solution ammonical to ascertain the completion of complexing (i.e. no precipitate should appear in the ammoniacal medium).

ix. If any precipitate appears, acidify the solution by adding more tartaric acid and ammonium hydroxide. The solution should be clear without any precipitation.

x. Add HCl until the solution is slightly acidic, heat to 60 to 80°C and then add 10 ml of 10% dimethyl glyoxime alcohol.

xi. Add ammonia solution till a faint smell is perceptible and then add a little excess.

xii. Keep the solution over hot water bath for 2 hours. Precipitation will start. Allow precipitation to continue till the next day.

xiii. Filter through a weighed sintered glass crucible.

xiv. Wash the precipitate with cold water.

xv. Dry the precipitate at 150°C and weigh (W_2).

$$Ni~(\%) = \frac{20.31 \times W_2}{W_1}$$

4.10. ESTIMATION OF Cu, Pb AND Zn IN POLYMETALLIC ORES

Polymetallic ores contain copper minerals like chalcopyrite ($CuFeS_2$), cuprite (Cu_2O), tenorite (CuO), chalcocite (Cu_2S), covellite (CuS), bornite (Cu_5FeS_4), malachite $[Cu_2CO_3(OH)_2]$, azurite $[Cu_3(CO_3)_2(OH)_2]$, etc. lead minerals like galena (PbS), cerussite ($PbCO_3$), anglesite ($PbSO_4$), etc. and zinc minerals such as sphalerite (ZnS), zincite (ZnO), smithsonite ($ZnCO_3$), hemimorphite $[Zn_4(Si_2O_7)(OH)_2 \times H_2O]$, etc. Because of their similar genesis, these ore minerals occur invariably in association with each other.

4.10.1. Preparation of Stock Solution

i. Take 1 g of the ore powder in a beaker, add 20 ml conc. HCl, and heat on a hot plate for 30 minutes.

ii. Add 10 ml conc. HNO_3 and keep over the hot plate till reaction ceases.

iii. Add 5 ml 1:1 H_2SO_4 and evaporate to dryness.

iv. Add 10 ml 1:1 H_2SO_4 to the resultant dry mass and make the volume to nearly 100 ml with addition of water.

v. Boil over a hot plate for about 15 minutes and filter thorough Whatman No. 40 filter paper.

vi. Wash the residue with hot water containing a few drops of H_2SO_4. Preserve the filtrate, which is the stock solution for estimation of Cu and Zn.

vii. Transfer the residue to the original beaker.

viii. Add 100 ml 5.5 pH mixture of ammonium acetate and acetic acid to the beaker, boil for 15 minutes and filter this through Whatman No. 40 filter paper.

ix. Wash the residue with hot amonium acetate and acetic acid buffer 3–4 times, filter and add the filtrate to that obtained at step (viii) above. This is the stock solution for estimation of Pb.

4.10.2. Estimation of Copper by Sodium Thiosulphate Titration Method

i. Take 50 ml aliquot for copper estimation obtained at step 4.10.1. (vi).

ii. Add about 20 ml dil. H_2SO_4, heat the mixture till all the fumes are expelled.

iii. Treat the mixture with dil. NH_3 to neutralize all mineral acids.

iv. Add 10 ml glacial acetic acid and 2.3 g of NaF.

v. Boil the mixture and cool in water bath.

vi. Add 2–3 ml KI and titrate the liberated iodine with 1 N $Na_2S_2O_3$ solution (24.82 gram $Na_2S_2O_3$ in 1 liter distilled water).

vii. Add starch near the end point. The colour will change from blue to colourless at the end point.

viii. Determine the amount of Cu by the following equivalence.

$$1~ml~1N~Na_2S_2O_3 = 0.06357~g~of~Cu$$

4.10.3. Estimation of Zinc by EDTA Complexometry Method

i. Take 50 ml aliquot obtained at step 4.10.1. (vi) in a 250 ml beaker and add dil. NH_3 (precipitate should not appear).

ii. Add 2 g of urea and boil vigorously, check the pH of the solution continuously adjust the pH to about 5.

iii. Remove the beaker from the flame and cool by cold water.

iv. Filter through Whatman No. 41 filter paper with little filter paper pulp.

v. Wash the residue 5–6 times with hot water. Collect the filtrate in a 400 ml beaker.

vi. Add 0.5–1.0 g ascorbic acid. Dissolve the solid by slow heating. Cool the beaker.

vii. Add 0.5–1 g thiourea and heat the beaker slowly till all solids are dissolved. Cool the beaker.

viii. Add 50 ml, 5.5 pH, sodium acetate buffer and 2–3 drops of xylenol orange indicator, which will impart purple red colour to the solution.

ix. Titrate the solution against standard disodium dihydrate salt of EDTA solution. At the end point purple red colour will change to golden yellow.

x. Determine the amount of Zn by the following equivalence.

$$1 \text{ ml lM EDTA} = 0.06537 \text{ g of Zn}$$

4.10.4. Estimation of Lead by EDTA Complexometry Method

i. Take 50 ml aliquot obtained at step 4.10.1 (ix) and add 5 ml dil. H_2SO_4.

ii. Wet the mass left out with dil. H_2SO_4 and add 50 ml water.

iii. Boil the mixture on burner. Cool and filter this through Whatman No. 40 filter paper.

iv. Wash the precipitate with hot water containing 5% H_2SO_4.

v. Transfer the filter paper along with precipitate to the original beaker.

vi. Add 100 ml, 5.5 pH, ammonium acetate-acetic acid buffer.

vii. Boil the mixture and cool in water bath.

viii. Add a pinch of ascorbic acid and a pinch of thiourea. Heat slowly to dissolve the solids.

ix. Add 25 ml sodium acetate 5.5 pH buffer and 2–3 drops of xylenol orange indicator. Titrate this against standard disodium dihydrate salt of EDTA solution. Colour will change from purple red to golden yellow at the end point.

$$1 \text{ ml lM EDTA} = 0.2072 \text{ g of Pb}$$

4.11. ESTIMATION OF TIN IN TIN ORE

Cassiterite (SnO_2) is the chief ore of tin. The procedures of preparation of necessary reagents and estimation of tin in cassiterite are as follows.

Reagents

a. *Standard 0.1 N iodine solution*: Dissolve 12.692 g of iodine and 26 g of KI in 400 ml water in a 500 ml beaker. Wash the contents into a 1 litre graduated flask and add distilled water to make the volume 1 litre. Determine the exact normality of this solution by a standard $Na_2S_2O_3$ solution.

b. *Starch solution*: Mix about 1 g of soluble starch with 20 ml cold water. Add the mixture slowly to 80 ml boiling water. Boil for few minutes.

i. Take 0.1 to 0.5 g of the finely ground sample in an iron or zirconium crucible, add about 8 times of its weight of sodium peroxide and mix well with a dry glass rod.

ii. Sprinkle a thin layer of peroxide over the mixture and heat gently over a low flame at a temperature just sufficient to produce complete fusion.

iii. Keep the melted mass at a dull redness for about 5 minutes and give the crucible 5–6 times swirling motions with the help of tongs.

iv. Remove the crucible from the heat and allow it to cool.

v. Submerge the crucible in about 100 ml water in a 500 ml beaker and cover the beaker with a watch glass to prevent loss due to spattering.

vi. After completion of disintegration of the fused mass remove the crucible and wash it with water.

vii. Add conc. HCl until the precipitate is completely dissolved.

viii. Filter the solution through Whatman No. 40 filter paper and wash the filter paper with hot water.

ix. Transfer the acid solution into a 500 ml conical flask, add 50 ml conc. HCl, 10 ml of 1:1 H_2SO_4 and 15 g granulated lead or lead foil.

x. Boil the solution for about an hour in the presence of CO_2 to reduce the Sn^{4+} to Sn^{2+} state.

xi. Remove the flask from the heat and cool rapidly in a cool water bath while maintaining a stream of CO_2. Add a few ml of starch solution and titrate immediately with 0.1 N iodine solution till a faint permanent blue tinge appears.

xii. Compute the percentage of tin from the following factor.

$$1 \text{ ml of } 1 \text{ N iodine} = 0.05935 \text{ g of Sn}$$

4.12. ESTIMATION OF TITANIUM IN TITANIUM ORE

Ilmenite ($FeO.TiO_2$) and rutile (TiO_2) are the principal sources of titanium. The methods of preparation of reagents and estimation of TiO_2 are given below.

Reagents

a. *Standard ferric ammonium sulphate* [$Fe(NH_4)(SO_4)_2 \cdot 12H_2O$] *solution*: Dissolve 60 g of ferric ammonium sulphate in 600 ml distilled water acidified with 20 ml 1:1 H_2SO_4. Add $KMnO_4$ solution till pink colour begins to appear. Dilute the solution to 2 litres and standardise against 0.1 N $K_2Cr_2O_7$ solution (2.94 gram of $K_2Cr_2O_7$ in 1 liter distilled water) to determine the exact normality.

b. *Ammonium thiocyanate (NH_4CNS) solution*: Dissolve 24 g of ammonium thiocyanate in 100 ml distilled water.

i. Take 0.2 g of sample powder in a silica crucible, add approximately 1 g of potassium pyrosulphate and fuse over a low flame till no black particle remains.

ii. Cool the crucible and dissolve the content in 1:1 H_2SO_4 and add water to make up to 250 ml in a volumetric flask. The acid strength in the volumetric flask should be approximately 5%.

iii. Take the solution in a wide-mouthed 500 ml conical flask with a 2-hole rubber stopper.

iv. Add 25 ml conc. HCl and nearly 2 g of aluminium foil to the flask.

v. Close the mouth of the flask with the rubber stopper and through one hole allow a slow stream of CO_2 from a Kipp's apparatus to pass into the flask. The other hole serves as an outlet.

vi. Boil the solution in the flask for more than half an hour to expel excess hydrogen gas. Maintain a slow stream of CO_2 throughout the boiling process.

vii. Cool the flask and remove the stopper and the delivery tube.

viii. Add 5 ml ammonium thiocyanate solution that serves as an indicator and titrate quickly with ferric ammonium sulphate [$FeNH_4(SO_4)_2.12H_2O$] solution to a faint red colour.

ix. Calculate TiO_2 from the following factor:

$$1 \text{ ml } 1 \text{ N } FeNH_4(SO_4)_2.12H_2O = 0.08 \text{ g } TiO_2 = 0.04795 \text{ g } Ti$$

4.13. ESTIMATION OF VANADIUM IN VANADIUM ORE

Important vanadium minerals are patronite (VS_4), vanadinite [$Pb_4(PbCl)(VO_4)_3$], carnotite ($K_2O.2U_2O_3.V_2O_5.3H_2O$) and titaniferous vanadium ores. The procedure of estimation of vanadium is as follows:

i. Fuse 0.2 to 0.5 g of finely ground ore with 5 times of its weight of sodium peroxide in an iron or nickel crucible.

ii. Transfer the product into a 500 ml beaker, add water and boil.

iii. Filter through Whatman No. 41 filter paper adding some pulp.

iv. Reject the residue and proceed with the filtrate (sodium vanadate).

v. Boil the solution for 30 minutes to expel H_2O_2.

vi. Transfer the solution into a 500 ml beaker, neutralise with 1:1 H_2SO_4 and add 4–6 ml of acid in excess.

vii. Add ferrous ammonium sulphate solution until a drop of the solution shows blue in a spot test with potassium ferricyanide solution.

viii. Add 10–15 ml H_3PO_4 and cool to room temperature.

ix. Add 0.1 N $KMnO_4$ until a light pink colour appears. Allow the solution to stand still for few minutes so that all vanadium is oxidised.

x. Destroy the excess $KMnO_4$ by 0.05 N $NaNO_2$ solution.

xi. Add 2 g of urea and stir. Allow it to stand for 5 minutes.

xii. Add 15 g of sodium acetate and stir to dissolve. If any precipitate is formed, add a drop of H_2SO_4 to clear it.

xiii. Add 8 drops of diphenylamine barium sulphonate indicator; allow 2–3 minutes for the development of full colour of the indicator.

xiv. Titrate with 0.1 N ferrous ammonium sulphate solution until the disappearance of blue colour.

xv. Calculate V or V_2O_5 from the following factors.

$$1 \text{ ml of } 1 \text{ N } Fe(SO_4).(NH_4)_2SO_4 = 0.05094 \text{ g } V = 0.09094 \text{ g } V_2O_5$$

4.14. ESTIMATION OF TUNGSTEN IN TUNGSTEN ORE

The important tungsten-bearing minerals are wolframite (FeMn)WO_4, scheelite (CaWO$_4$), tungstenite (WS_2), tungstite (WO_3), etc. Estimation of WO_3 is made by gravimetric analysis by the procedure given below.

i. Take 0.1 to 0.5 g (W_1) of the sample powder and add 70 ml conc. HCl and digest over a hot plate for about 3 hours at a low temperature till all black particles are dissolved.

ii. Add 20 ml conc. HNO_3 to the hot boiling solution, digest and evaporate to about 50 ml.

iii. Make the volume nearly 100 ml with hot water. Stir thoroughly and gently.

iv. Add 10 ml cinchonine hydrochloride solution (10 g of cinchonine dissolved in 100 ml of 6N HCl) and boil over the hot plate.

v. Allow the precipitate to settle and filter through Whatman No. 40 filter paper adding a little pulp.

vi. Wash the precipitate 3 times with 0.1% cinchonine hydrochloride solution.

vii. Dissolve the precipitate in the same beaker in hot conc. ammonia to convert all the tungsten acid to ammonium tungstate.

viii. Filter through Whatman No. 41 filter paper and wash the residue with hot water.

ix. Boil the filtrate until all free ammonia is expelled and dilute the solution to 100 ml with hot water.

x. Add 50 ml conc. HCl and 25 ml conc. HNO_3 to the boiling solution and continue boiling for some time.

xi. Add 10 ml cinchonine hydrochloride solution and 50 ml hot water and boil again.

xii. Allow it to cool to room temperature. Filter through Whatman No. 40 filter paper and wash the residue.

xiii. Ignite the precipitate in a weighed platinum crucible. The temperature should not exceed 750°C as WO_3 is volatile above this temperature.

xiv. Cool the ignited oxide and moisten with a few drops of water.

xv. Pour a few drops of HF and keep over a low heat to remove silica.

xvi. Heat to a temperature not exceeding 750°C. The dry residue should be of lemon yellow colour. Measure the weight of the residue (W_2).

$$WO_3(\%) = \frac{W_2 \times 100}{W_1} \text{ and } W\% = 0.79297 \times WO_3\%$$

4.15. ESTIMATION OF ZIRCONIUM IN ZIRCON

Important zirconium minerals are zircon ($ZrSiO_4$), eudialyte (Zr,Fe,Ca,Na-silicate), baddeleyite (ZrO_2), guarinite (Na_2,Ca)(Si,Zr)O_3, etc. The procedure of estimation of zirconium in zircon is given below.

i. Take about 1 g of zircon sample powder and 10 times of its weight of sodium peroxide in a platinum crucible.

ii. Heat the mixture on a burner till it becomes red and then heat in a muffle furnace at 1000°C for 15–30 minutes.

iii. Add HCl, cover the crucible with a watch glass and heat on water bath.

iv. When CO_2 evolution stops, remove the watch glass and evaporate the solution to dryness.

v. Repeat the process twice and finally dissolve the residue.

vi. Filter through sintered glass crucible. Wash the precipitate with NH_4Cl.

vii. Collect the filtrate and washings in 50 or 100 ml volumetric flask. The acidity of the solution should be 2N.

viii. Add a little Eriochrome Black T to the solution, boil for 7–10 minutes until a persistent blue–violet colour develops.

ix. Titrate the solution with 0.01 M EDTA (3.723 gram disodium dihydrate salt of EDTA in 1 liter water) until the colour of the solution turns violet-pink.

x. Heat the solution again and if blue violet colour is restored, continue titration with EDTA until the violet-pink colour persists.

xi. Calculate the Zr content by the following conversion factor.

$$1 \text{ ml of } 0.01 \text{ M EDT A} = 0.9122 \text{ mg of Zr}$$

4.16. ESTIMATION OF BaSO$_4$ IN BARIUM ORE

The chief mineral of barium is barite (BaSO$_4$) though witherite (BaCO$_3$) contains appreciable amount of Ba. In addition to BaSO$_4$, barite contains matters soluble in water, SO$_3$, Fe$_2$O$_3$, Al$_2$O$_3$, CaO, MgO and volatile matter. The procedure for estimation of BaSO$_4$ in barite is as follows:

i. Take about 0.2 g (W$_1$) of finely ground sample, 1.5 g of Na$_2$CO$_3$ and 1 g of K$_2$CO$_3$ in a platinum crucible.

ii. Cover the crucible and fuse the mixture by heating in a furnace at 1000°C for about 30 minutes.

iii. Take the crucible out of the furnace and rotate it so that the fusion will solidify in a thin layer by the side of the crucible. This will shorten the time required for leaching.

iv. When cooled, leach the fusion with 200 ml boiling water in a 500 ml beaker.

v. Filter the solution through Whatman No. 40 filter paper. Wash the filter paper and residue 10 to 12 times with hot Na$_2$CO$_3$ solution and finally with hot water till it is free from sulphate. (The filtrate may be preserved for the estimation of SO$_3$).

vi. Dissolve the residue from the filter paper in hot 1:1 HCl. Add 1 g NH$_4$Cl, and neutralize with NH$_4$OH using methyl red indicator. Boil and filter through Whatman No. 41 filter paper. Preserve the filtrate.

vii. Wash the residue 4 or 5 times with hot water, dissolve in hot HCl and reprecipitate with NH$_4$OH as before. Wash the residue with hot water 5 to 6 times. This operation will bring out any barium that might have been occluded in the R$_2$O$_3$ precipitate.

viii. Mix both the filtrates, take in a 500 ml beaker, add a few drops of methyl red indicator and neutralise the solution with HCl. Add 2 ml more conc. HCl.

ix. Dilute the solution to 400 ml with distilled water, boil and add 20 ml 10% hot (NH$_4$)$_2$SO$_4$ solution drop wise with constant stirring. Boil the solution for 5 minutes. Remove from the flame and allow it to stand for at least 4 hours or preferably overnight.

x. Filter the solution through Whatman No. 42 filter paper and wash the residue thoroughly with warm water containing 5 drops of H$_2$SO$_4$ per litre followed by cold water until the filtrate is free from chloride ions.

xi. Transfer the residue along with the filter paper to a weighed platinum crucible (W$_2$), slowly burn the filter paper and ignite the residue at 850° to 900°C for 30 minutes.

xii. Cool the crucible in a desiccator and weigh (W$_3$).

The percentage of BaSO$_4$ is calculated as follows:

$$BaSO_4 \ (\%) = \frac{W_3 - W_2}{W_1} \times 100$$

where, W_1 = weight of the sample taken
W_2 = weight of the empty platinum crucible
W_3 = weight of the platinum crucible and the ignited BaSO$_4$

4.17. ESTIMATION OF CaF$_2$ IN FLUORITE

Digest finely powdered fluorite sample with 8% AlCl$_3$ solution and 0.1 N HCl. After appropriate corrections made for all acid soluble contents of sample the loss in weight corresponds to CaF$_2$. The procedure of estimation of CaF$_2$ involves two parts (a) loss of weight in digestion and (b) correction for acid soluble matter.

4.17.1. Estimation of Loss of Weight in Digestion

i. Take 0.5 g (W_1) of powdered fluorite sample in a 250 ml beaker and add 25 ml AlCl$_3$ solution to it. Stir the mixture well with a glass rod.
ii. Digest the mixture on a steam bath for 1 hr.
iii. Filter the sample through Whatman No. 40 filter paper.
iv. Wash the residue 8–10 times with hot water.
v. Transfer the filter paper and residue into a tarred platinum crucible.
vi. Keep the crucible in a furnace and ignite the filter paper at 500° ±10°C.
vii. Take the crucible out of furnace, cool to room temperature in a desiccator and weigh the residue (W_2).

$$\text{Loss in weight (A) in \%} = \frac{W_1 - W_2}{W_1} \times 100$$

4.17.2. Estimation of Correction for Acid Soluble Matter

i. Take 0.5 g (W_3) of powdered fluorite sample in a 250 ml beaker and add 25 ml acetic acid to it.
ii. Digest the mixture on a steam bath for 1 hour.
iii. Filter the sample through Whatman filter paper No. 40 and wash the residue 8–10 times with hot water.
iv. Keep the filter paper and residue in a tarred platinum crucible.
v. Transfer the crucible to a furnace and ignite the filter paper at 500° ±10°C.
vi. Take the crucible out of furnace; cool to room temperature in a desiccator and weigh the residue (W_4).

$$\text{Acid soluble matter (B) in \%} = \frac{W_3 - W_4}{W_3} \times 100$$

$$\text{CaF}_2 \ \% \text{ in sample} = A - B$$

4.18. ESTIMATION OF CaSO$_4$.2H$_2$O IN GYPSUM

i. Take about 1 g (W_1) of sample in a 500 ml beaker and add 100 ml of 10% alkaline ammonium acetate solution.
ii. Digest the mixture on a steam bath for 1 hour.

iii. Filter the sample through sintered glass crucible of porosity G3 under suction.

iv. Transfer the residue into a crucible and wash 5–6 times with cold water.

v. Heat the crucible to a constant weight at 100°C.

vi. Determine the weight of the residue (W_2).

$$CaSO_4.2H_2O \text{ (in \%)} = \frac{W_1 - W_2}{W_1} \times 100$$

4.19. ESTIMATION CaO AND P_2O_5 IN APATITE

4.19.1. Preparation of Stock Solution

i. Take about 1 g (W_1) of apatite sample in a beaker and add 20 ml of conc. HCl and 5 ml conc. HNO_3 to it.

ii. Heat the mixture over a hot plate till the sample is digested and acids evaporate to dryness.

iii. Add about 50 ml of water to the beaker and boil the content for 5 minutes.

iv. Cool the beaker and filter the content through Whatman No. 40 filter paper.

v. Preserve the filtrate and wash the residue 5–6 times with hot water.

vi. Take the residue along with the filter paper in a platinum crucible, burn the filter paper and ignite the residue.

vii. Cool the residue in a desiccator to room temperature and weigh (W_2).

viii. Moisten the residue with 5 drops of 1:1 H_2SO_4 and add 10 ml HF.

ix. Evaporate the liquid completely and ignite the residue till it is red hot.

x. Cool the residue in a desiccator and weigh (W_3).

xi. The loss of weight, i.e.

$$\text{Percentage of silica} = \frac{W_2 - W_3}{W_1} \times 100$$

xii. Fuse the residue with Na_2CO_3 and mix with the filtrate obtained at step (v) above.

xiii. Take the filtrate in a 250 ml volumetric flask and make up to the volume by adding distilled water.

4.19.2. Estimation of Total CaO

i. Take 25 ml of aliquot in a beaker, add 0.05 g of citric acid and boil the mixture.

ii. Add a few drops of bromophenol (blue indicator) and about 25 ml saturated ammonium oxalate solution till the solution develops a dirty blue colour.

iii. Boil the solution till precipitation starts and then keep the solution out of flame for at least 2 hours.

iv. Filter the content through Whatman No.40 filter paper and wash the residue with hot water till it is free from oxalate ions, which can be tested with $KMnO_4$ and 1:1 H_2SO_4.

v. Dissolve the residue in 10 ml 1:1 H_2SO_4, add 50 ml demineralised water and heat to 75°C for about 5 minutes.

vi. Cool the solution and titrate with 0.1 N $KMnO_4$ solution till a permanent pink colour develops.

$$\text{1 ml 1 N } KMnO_4 = 0.02804 \text{ g of CaO}$$

4.19.3. Estimation of Free CaO

i. Take 0.5 g of powdered sample in a volumetric flask, add 25 g of cane sugar and 200 ml of distilled water.

ii. Shake the mixture well and allow it to remain as such for about 2 hours.

iii. Titrate 25 ml of the solution against 0.1 N oxalic acid using phenolphthalein indicator.

$$1 \text{ ml } 1 \text{ N oxalic acid} = 0.025 \text{ g of CaO}$$

4.19.4. Estimation of P_2O_5

i. Take about 1 g of powdered sample in a 500 ml conical flask, add 10 ml conc. HCl and heat over a burner till the powder goes into solution.

ii. Add 2 ml conc. HNO_3 and heat the mixture for 10 minutes.

iii. Neutralise the solution with NaOH solution till a faint precipitate appears.

iv. Add 2 ml of conc. HCl and water to make the volume 120 ml.

v. Add 0.5 g citric acid to the solution and heat till boiling.

vi. Add 30 ml of citromolybdate solution and heat till boiling.

vii. Add 25 ml of quinoline hydrochloride solution slowly by a burette and boil the solution again for 10 minutes. Precipitation will be initiated.

viii. Allow the precipitate to settle for at least 3 hours or preferably overnight.

ix. Filter the precipitate through a pad of paper pulp under suction.

x. Wash the precipitate with cold water till it is free from acid (to be tested by litmus paper).

xi. Transfer the precipitate to the original flask and add 25, 50, 75 or 100 ml of 0.1 N NaOH solution till the precipitate is dissolved completely.

xii. Add 5–6 drops of thymol blue or phenolphthalein indicator to the solution and titrate the excess of alkali against 0.1 N HCl.

xiii. Titrate 25 ml of NaOH with 0.1 N HCl, the value of which is to be used to calculate the exact volume of NaOH used up by the precipitate.

$$1 \text{ ml of } 1 \text{ N HCl} = 0.001193 \text{ g of P} = 0.002733 \text{ g of } P_2O_5$$

Analysis of Coal

Analysis of coal is divided into three parts, namely proximate analysis, ultimate analysis and analysis of ash. Proximate analysis gives the rough grade of the coal whereas ultimate analysis and estimation of calorific value give a closer look at the quality of coal. Analysis of ash gives the composition of combustion waste. According to the Indian standard, the gross sample is collected at different levels from bulk of coal in heaps. The sample is crushed to less than 6 mm size and about 1.2 kg of the powdered coal is taken by coning and quartering, which is preserved in a stoppered bottle for analysis.

5.1. PROXIMATE ANALYSIS

Proximate analysis includes estimation of amounts of moisture, ash, volatile matter and fixed carbon, out of which first three components are estimated in the laboratory and fixed carbon content is computed by subtracting the sum of moisture, ash and volatile matter from 100.

5.1.1. Estimation of Moisture (M) in Coal

The moisture of coal comprises inherent moisture and surface moisture. The first type is associated with coal at the time of its mining while the second type is acquired after mining. Weight lost by coal by heating it for 1 hour in an air oven at 105°C is commonly regarded as the moisture content of the coal. This is, however inaccurate as the air contains some amount of moisture. Use of current of nitrogen instead of air gives appreciably good result. Hence, the moisture estimation should always be carried out in a current of nitrogen. The procedure for estimation of moisture is as follows:

   i. Take about 1 g (W_1) of coal sample in a silica patty dish.
   ii. Tap the dish to spread the sample uniformly in a thin layer.
   iii. Place the patty dish in a muffle furnace maintained at 105°C in a current of nitrogen for 1 hour.
   iv. Take the patty dish out of muffle furnace; cool it in a desiccator and weigh. Say the weight is W_2.
   v. The loss in the weight ($W_1 - W_2$) is the quantity of moisture in the coal sampl .

$$\text{Moisture (\%)} = \frac{W_1 - W_2}{W_1} \times 100$$

5.1.2. Estimation of Ash (A) in Coal

The residue left after coal burning is known as ash. The amount of ash varies considerably in different coals and in the coal of the same rank. The terms mineral matter and ash are not synonymous. Mineral matter refers to the various impurities present in coal, whereas ash is the solid residue that remains after the coal is completely burnt. Though ash is derived from mineral matter, it has not of the same composition and amount as the mineral matter. Roughly, mineral matter in coal is 10% more than the actual ash content. The procedure of estimation of ash in coal is as given below.

    i. Spread about 2 g (W_1) of sample in silica patty dish.

    ii. Place the patty dish in a furnace maintained at 500°C for 30 minutes and at 750 ± 50°C for further 60 minutes.

    iii. Cool the patty dish on a metal plate for 5 minutes and finally cool it to room temperature in a desiccator and weigh. The weight (W_2) of the left out residue is the ash.

$$Ash\ (\%) = \frac{W_2}{W_1} \times 100$$

The ash can be further chemically analysed in terms of oxide constituents. The procedure of estimation of SiO_2 and preparation of stock solution necessary for quantitative estimation of other oxides including rare and trace elements is given below.

    i. Fuse 1 g of ash with 3–4 g of sodium carbonate in a platinum crucible.

    ii. Cool the crucible and extract the mass in dilute HCl.

    iii. Evaporate the solution on a hot plate.

    iv. Bake this to dehydrate completely on hot plate for 30 minutes.

    v. Add 5 ml conc. HCl and 50 ml demineralised water.

    vi. Boil the mixture, cool and filter this through Whatman No. 40 filter paper.

    vii. Wash the residue 5–6 times with hot water.

    viii. Preserve the filtrate for further use.

    ix. Transfer the residue along with the filter paper to a platinum crucible.

    x. Heat the platinum crucible in a muffle furnace at 1000°C.

    xi. Cool, and weigh (W_1) the mass.

    xii. Add 2–3 drops of dil. H_2SO_4 and 10–15 ml of 48% HF.

    xiii. Heat the mixture on asbestos sheet to expel fumes from the crucible.

    xiv. Heat the crucible initially on a low temperature burner and finally heat at 1000°C.

    xv. Cool and weigh (W_2) the residue.

    xvi. The loss ($W_1 - W_2$) corresponds to the quantity of silica in the ash.

    xvii. Fuse the mass left after hydrofluorisation in the crucible with potassium pyrosulphate.

    xviii. Mix the fused mass with the filtrate preserved at step viii for further analysis.

Make up the filtrate to 250 ml in a volumetric flask. This is the stock solution, which can be used for estimation of different oxides and elements present in coal ash.

5.1.3. Estimation of Volatile Matter (VM) in Coal

Loss of weight of the coal sample at 930 ± 20°C is regarded as volatile matter, which is composed of CO, CO_2, H_2, N_2, O_2, CH_4, coal gas, ammonia gas, benzene, toluene, phenol and tar. Though many of them are combustible, their heat energy is rarely utilised in any furnace. These are either driven off with waste gases or they burn at the mouth of furnaces and boilers. Thus, the energy released is of little use. The procedure of estimation of volatile matter in coal is as follows:

i. Take 1 g of coal sample (preferably dried at 105°C) in volatile matter crucible.

ii. Tap the crucible to spread the sample uniformly in a thin layer.

iii. Add 2–3 drops of benzene to the coal sample and place the crucible in a muffle furnace maintained at 930 ± 20°C for 7 minutes.

iv. Cool the crucible in a desiccator and weigh.

v. The loss corresponds to the quantity of volatile matter. Express it in percent.

5.1.4. Estimation of Fixed Carbon (FC)

Fixed carbon is not determined directly. It is the difference between the sum of other components and 100.

$$Fixed\ carbon\ (FC) = [100 - (M\% + A\% + VM\%)]$$

5.1.5. Reporting Proximate Analysis Values

Different terminologies are used in reporting proximate analysis values. The terminologies along with their calculations are given below.

Different bases are given in following equations and illustrated in diagrammatic form in Fig. 5.1.

a. As received coal = FC + VM + A + M

b. Air-dried coal = FC + VM + A + M (inherent)

c. Dry coal = FC + VM + A

d. Dry ash free coal = FC + VM

e. Dry mineral matter free coal = FC + VM (organic)

a. As Received Basis

When the consumer receives the coal from the mine, its analysis is reported as *as received basis*. The results of the proximate analysis are expressed as percentages of the coal including the total moisture content.

	Fixed carbon	Volatile matter		Ash	Moisture	
		Organic	Mineral		Inherent	Acquired
(e)	←——Dry mineral matter free——→					
(d)	←——Dry ash free——————————→					
(c)	←——Dry——————————————————→					
(b)	←——Air dried————————————————————→					
(a)	←——As received——————————————————————————→					

Fig. 5.1: Diagrammatic representation of different coal types on the basis of proximate analysis

i. Moisture (M) % = $\dfrac{\text{Weight of moisture}}{\text{Weight of the coal sample as received}} \times 100$

ii. Ash (A) % = $\dfrac{\text{Weight of ash}}{\text{Weight of the coal sample as received}} \times 100$

iii. Volatile matter (VM)% = $\dfrac{\text{Weight of VM}}{\text{Weight of the coal sample as received}} \times 100$

iv. C % = [100 – (M% + A% + VM%)]

b. Air-dried Basis

The coal is said to be *air dried* when it is exposed to an artificial and standard atmosphere at 40°C and 60% relative humidity. In the process, the acquired or surface moisture is lost, but the coal still contains some amount of inherent moisture. The proximate analysis results are expressed as percentages of air-dried coal, including inherent but not surface or free moisture.

i. Moisture (M) % = $\dfrac{\text{Weight of moisture}}{\text{Weight of the air dried coal sample}} \times 100$

ii. Ash (A) % = $\dfrac{\text{Weight of ash}}{\text{Weight of the air dried coal sample}} \times 100$

iii. Volatile matter (VM) % = $\dfrac{\text{Weight of VM}}{\text{Weight of the air dried coal sample}} \times 100$

iv. FC % = [100 – (M% + A% + VM%)]

c. Dry or Moisture Free

When the effect of moisture is completely eliminated from the analytical data, the coal analysis is termed as *dry* or *moisture free* basis.

i. Ash (A) % = $\dfrac{A}{(100 - M)} \times 100$

ii. Volatile matter (VM) % = $\dfrac{VM}{(100 - M)} \times 100$

iii. FC % = $\dfrac{FC}{(100 - M)} \times 100$

d. Dry Ash Free (Daf)

Data generated by eliminating the effect of moisture and ash is termed as *dry ash free* (daf) basis. This data is suitable for comparing low ash (<10%) coals.

i. Volatile matter (VM) % = $\dfrac{VM}{(100 - M - A)} \times 100$

ii. FC % = $\dfrac{FC}{(100 - M - A)} \times 100$

e. Dry Mineral Matter Free (Dmmf)

In high ash coals, the mineral matter is about 10% more than its ash while in case of low ash coals, the mineral matter is nearly equal to its ash content. Thus, the data expressed on *dry mineral matter free* basis is most suitable for comparing high ash coals, in which ash content exceeds 10%.

Mineral matter (MM) % = A + 0.1 A = 1.1 A

i. Volatile matter (VM) % = $\dfrac{(VM - 0.1\,A)}{(100 - M - 1.1 - A)} \times 100$

ii. FC % = $\dfrac{FC}{(100 - M - 1.1 - A)} \times 100$

5.2. ULTIMATE ANALYSIS

Ultimate analysis of coal refers to estimation of percentages of elements like carbon, hydrogen, nitrogen, sulphur and oxygen present in dry mineral matter free coal. The first four elements can be determined directly, but there is no satisfactory method for estimating oxygen. Thus, the amount of oxygen present in coal is found out by subtracting the total percentages of carbon, hydrogen, nitrogen and sulphur from 100.

5.2.1. Estimation of Carbon and Hydrogen

A known amount of coal is burnt in a current of oxygen, as a result of which carbon reacts with oxygen to form CO_2 and hydrogen reacts with oxygen to form H_2O. The generated CO_2 and H_2O are passed through weighed tubes of anhydrous KOH and $CaCl_2$, respectively, where they are absorbed. The increase in the weight of KOH tube represents the weight of CO_2 while the increase in the weight of $CaCl_2$ tube represents the weight of H_2O.

Calculation

Let weight of the coal sample taken = W_1

Increase in weight of KOH = W_2 = amount of CO_2 formed

Increase in weight of $CaCl_2$ = W_3 = amount of H_2O formed

From the equation $C + O_2 = CO_2$ it is apparent that 12 g of C and 32 g of O form 44 g of CO_2, i.e. 44 g of CO_2 contains 12 g of C

$\Rightarrow 1$ g of CO_2 contains $\dfrac{12}{44}$ g of C

$\Rightarrow W_2$ g of CO_2 contains $\dfrac{12}{44} \times W_2$ g of C

W_1 g of coal contains $\dfrac{12}{44} \times W_2$ g of C

$\Rightarrow 1$ g of coal contains $\left(\dfrac{12}{44} \times W_2\right) \div W_1$ g of C

$\Rightarrow 100$ g of coal contains $\left[\left\{\dfrac{12}{44} \times W_2\right\} \div W_1\right] \times 100$ g of C

$\therefore$ Percentage of carbon (C) in coal $= \dfrac{12 \times W_2}{44 \times W_1} \times 100 = \dfrac{300\, W_2}{11\, W_1}$

From the equation $2H_2 + O_2 = 2H_2O$ it is apparent that 4 g of H and 32 g of O form 36 g of H_2O, i.e. 9 g of H_2O contains 1 g of H

$\Rightarrow$ 1 g of H_2O contains $(1 \div 9)$ g of H

$\Rightarrow W_3$ g of H_2O contains $\dfrac{W_3}{9}$ g of H $\Rightarrow W_1$ g of coal contains $\dfrac{W_3}{9}$ g of H

$\Rightarrow$ 1 g of coal contains $\dfrac{W_3}{9\, W_1}$ g of H $\Rightarrow$ 100 g of coal contains $\dfrac{100\, W_3}{9\, W_1}$ g of H

$\therefore$ Percentage of hydrogen (H) in coal $= \dfrac{100\, W_3}{9\, W_1}$

5.2.2. Estimation of Nitrogen

There are two methods for estimation of nitrogen content of coal, both being known as Kjeldahl's method.

a. Method 1

   i. Weigh accurately 0.1 to 0.2 g of the sample and 2 g of catalyst (87 parts of K_2SO_4 and 13 parts of $HgSO_4$) and put these in a Kjeldahl flask.

   ii. Shake the flask well to mix the coal and catalyst thoroughly. Add 10 ml conc. H_2SO_4 and shake well.

   iii. Heat the flask by a Bunsen burner keeping it on a hole of an asbestos board (Fig.3.4). The hole should cover the liquid. The neck of the flask should be inclined at an angle of 60°.

   iv. Heat gradually to boiling. Boil for about 15 to 20 minutes and add a small volume of conc. H_2SO_4, if required.

   v. Cool the solution, dilute with about 200 ml of water, add few pieces of cracked porcelain to prevent from bumping and add 25 ml of solution containing 80 g of sodium thiosulphate or 40 g of sodium or potassium sulphide per litre.

   vi. Add sufficient quantity of NaOH solution to make the solution strongly alkaline.

   vii. Connect the condenser.

   viii. Place the other end of the condenser dipped in measured solution of standard H_2SO_4.

   ix. Distill about 150–200 ml into the standard H_2SO_4 solution.

   x. Remove the dipped end of the condenser.

   xi. Put off the flame, disconnect the flask and wash the condenser by standard H_2SO_4.

   xii. Add 2–3 drops of methyl red indicator.

   xiii. Titrate this against standard alkali solution.

   xiv. Estimate the quantity of nitrogen in the sample by the following equivalence.

$$\text{1 ml 0.5 N } H_2SO_4 = 0.007 \text{ g of nitrogen}$$

Express it by percent weight.

b. Method 2

i. Weigh accurately 0.1 to 0.2 g of the coal sample and add 2 g of catalyst (K_2SO_4 and $CuSO_4$ mixture). Transfer the mixture into a Kjeldahl flask (*see* Fig. 3.4).

ii. Shake the flask well to mix the content thoroughly, add 10 ml conc. H_2SO_4 and shake well.

iii. Heat the mixture gradually to boiling; continue boiling for 15 to 20 minutes. Nitrogen of coal will react with H_2SO_4 forming $(NH_4)_2SO_4$. When the total amount of nitrogen is converted to $(NH_4)_2SO_4$, the colour of the solution will become clear. Add small volume of conc. H_2SO_4 if required.

iv. The clear solution is filtered with Whatman No. 40 filter paper and treated with 50% NaOH solution.

v. The ammonia thus formed is distilled over and absorbed in a known quantity of standard H_2SO_4.

vi. The volume of unused H_2SO_4 is determined by titrating against standard NaOH solution and the amount of H_2SO_4 neutralised by ammonia (liberated from coal) is found out.

vii. Percent of nitrogen in coal is determined from following equivalence.

$$\text{Nitrogen } \% = \frac{\text{Volume of acid used} \times \text{Normality} \times 100}{\text{Weight of coal taken}}$$

5.2.3. Estimation of Sulphur in coal by Eschka's Method

i. Mix thoroughly 1 g of coal with 3 g of Eschka mixture (porous calcined magnesia and anhydrous sodium carbonate in 2:1 proportion).

ii. Take the mixture in a platinum crucible and cover it by about 2 g of Eschka mixture.

iii. Fuse the mass carefully at a low flame till all black particles burn out.

iv. Cool the crucible. Extract the mass in 100 ml demineralised water in a 250 ml beaker by gradual heating on hot plate.

v. Remove crucible from the hot plate and filter the content through Whatman No. 40 filter paper.

vi. Wash the residue 5–6 times with hot water.

vii. Add 10 ml bromine water and acidify the filtrate with dilute HCl.

viii. Boil the mixture and add 20 ml of 10% $BaCl_2$ solution.

ix. Boil for 30 minutes. Sulphur present in coal comes out as sulphate, which reacts with $BaCl_2$ to form $BaSO_4$.

x. Cool this for 3–4 hours preferably overnight.

xi. Filter the precipitate through Whatman No. 42 filter paper.

xii. Wash the precipitate 5–6 times with cold water.

xiii. Ignite the filter paper along with the precipitate in a platinum crucible at 800°C.

xiv. Cool the crucible and determine the weight of $BaSO_4$.

$$S\% = \frac{(13.74 \times \text{wt. of BaSO}_4)}{\text{Wt. of the coal sample taken}}$$

5.2.4. Estimation of Oxygen in Coal

Oxygen percent in coal = [100 − (C % + H % + N % + S %)] in dry mineral matter free coal. If the coal is not free from mineral matter, the percentage of ash is to be subtracted from the result of the above equation.

5.3. ESTIMATION OF CALORIFIC VALUE (CV) OF COAL

Calorific value (CV) of coal is the amount of heat generated by complete combustion of unit weight of coal. The units of CV are calories/gram (cal/g), kilo calories/kg (kcal/kg) and Btu/lb in CGS, MKS and FPS systems, respectively. 1 calorie is the amount of heat required to raise the temperature of 1 g of water from 14.5°C to 15.5°C. Btu is the amount of heat required to raise the temperature of 1 lb water from 60 to 61°F. 1 kcal/kg is equal to 1.8 Btu/lb.

The calorific value (CV) of coal is determined by high-pressure bomb calorimeter that consists of a strong stainless steel vessel, known as bomb. It can withstand high pressure. The bomb has a lid, which be screwed firmly on the bomb. The lid is provided with two electrodes, one of which has a ring to hold the silica crucible. There is an inlet valve through which oxygen can enter into the bomb.

The bomb is placed in a copper calorimeter with a known weight of water. An air jacket and one water jacket surround the calorimeter to prevent loss of heat by radiation. The calorimeter has an electrical stirrer for starring water and a Beckmann thermometer that records the temperature of water.

A known weight (W_1 kg) of powdered coal sample is taken in the silica crucible, which is supported over the ring of the calorimeter. A fine magnesium wire is stretched across the electrodes touching the coal sample. Oxygen is forced into the bomb till the pressure becomes 25–30 atmospheres. Initial temperature (T_1°C) and weight W_2 kg of water are noted. The current is switched on, as a result of which, the coal in the crucible burns. The heat produced by burning of coal is absorbed in water, which is stirred continuously by the electric stirrer. The thermometer records the maximum temperature (T_2°C) of the water. The calorific value (CV) of the coal is computed as follows:

Weight of the coal taken in crucible = W_1 kg

Weight of the water in calorimeter = W_2 kg

Water equivalent of calorimeter, stirrer and bomb = W_3 kg

Initial temperature of water in calorimeter = T_1°C

Final temperature of water in calorimeter = T_2°C

Let the calorific value of coal is CV kcal/kg

Heat gained by water = $W_2 (T_2 - T_1)$ kcal

Heat gained by calorimeter = $W_3 (T_2 - T_1)$ kcal

Total heat gained = $W_2 (T_2 - T_1) + W_3 (T_2 - T_1)$ kcal = $(W_2 + W_3)(T_2 - T_1)$ kcal

Heat liberated by W_1 kg of coal = $W_1 \times CV$ kcal/Kg

Since heat liberated by coal = heat gained by water and calorimeter

$$W_1 \times CV \text{ kcal/Kg} = (W_2 + W_3)(T_2 - T_1) \text{ kcal}$$

$$\therefore \text{Calorific value of coal (CV)} = \frac{(W_2 + W_3) \times (T_2 + T_1)}{W_1} \text{ kcal/kg}$$

In addition to the experimental estimation, there are a number of formulae for estimation of calorific value of coal. These are:

i. Seyler's formula: $CV = 388.1\,H + 123.92\,C + \tfrac{1}{4}\,O^2 - 4269$

ii. Goutal's formula: $CV = 82\,FC + \alpha\,VM$

α is a constant related to VM % (on dry ash free basis) by the following relation:
$$\alpha = 160 - 2.74\,VM - 0.0823\,VM^2 + 0.0065\,VM^3 - 0.0001\,VM^4$$
with goodness of fit 99.89%

iii. Dulong's formula: $CV = 80.8\,C + 344 \times \left(H - \dfrac{O}{8}\right)$, when O >10%

$$CV = 337\,C + 1442 \times \left(H - \dfrac{O}{8}\right) + 93\,S, \text{when O} <10\%$$

iv. Mazumdar's formula: $CV = 0.512\,(C + H) + 81.07 \times \left(\dfrac{H}{C}\right) - 16.46$ [CV in MJ/kg]

v. Formulae developed by Central Fuel Research Institute (CFRI):

$CV = 91.7\,FC + 75.6\,(VM - 0.1\,A) - 60\,M$, when M $\leq$2%

$CV = 85.6\,[100 - (1.1\,A + M)] - 60\,M$, when M >2%

where,

M = Moisture in %	C = Carbon in %
A = Ash in %	H = Hydrogen in %
VM = Volatile matter in %	O = Oxygen in %
FC = Fixed carbon in %	S = Sulphur in %

Analysis of Water Sample

Water is essential for sustenance of life in the earth. It is used for drinking, irrigation and in industries. It is available in form of surface water and groundwater. Water is regarded as a universal solvent and the salts present in it occur in dissociated state of cations and anions. The quantities of different chemical constituents determine its suitability or otherwise in a particular field of use. The important constituents are calcium, magnesium, sodium, potassium, carbonate, bicarbonate, chloride and sulphate in addition to nitrate, phosphate, fluoride, iron, manganese, aluminium, silica, etc. Parameters associated with water chemistry are pH, electrical conductivity (EC), total dissolved solids (TDS), total hardness (TH), total alkalinity (TA), turbidity, etc.

6.1. INSTRUMENTS FOR WATER ANALYSIS

Nowadays, ion meter, pH meter, water analyzer kit and turbidity meter are available with the help of which parameters like temperature, pH, conductance, total dissolved solids, salinity and turbidity of the water samples can be determined conveniently. Quantitative estimation of many of the chemical constituents can be determined by gravimetric and titrimetric methods. However, some of the constituents, which are present in lesser quantities, necessitate the use of instruments like flame- and spectro-photometers. The operating procedures of the flame photometer have been described in sections 3.8.4 and 3.8.5. The general operating procedures of ion/pH meter, spectrophotometer and water analyzer are briefly described below.

6.1.1. Ion/ pH Meter

The instrument measures the ionic concentration of elements in a solution using individual ion selective electrodes (ISE). The pH of the solution can be measured using a pH electrode, which measures the concentration of H^+ ions. Basically pH electrode and ion selective electrodes are of the same family. A pH electrode is an ion selective electrode which is specifically sensitive to H^+. The ion-meter can measure the ionic concentration of ammonia (NH^{3+}), bromide (Br^-), cadmium (Cd^{2+}), calcium (Ca^{2+}), chloride (Cl^-), copper (Cu^{2+}), cyanide (CN^-), fluoride (F^-), iodide (I^-), lead (Pb^{3+}), nitrate (NO_3^-), nitrite (NO_2^-), potassium (K^+), silver (Ag^+), sodium (Na^+), sulphide (S^{2-}), etc. The precision of the instrument ranges from 0.01 ppm to 10,000 ppm with relative accuracy of ± 0.5% of the reading. The pH meter is capable of measuring pH from 0–14, with a resolution of 0.001 pH.

a. General Features

The block diagram of ion/pH meter model 363 of systronics make is shown in Fig. 6.1. The keyboard, channel enunciator and display units are present in the front panel. The keyboard has 12 numeric and 11 functional keys. The channel enunciator shows three channels, 1, 2 and difference (2–1). The display unit has 2 lines 20-character LCD to interact with the user while setting up and programming of the instrument as well as to present the end results. Temperature probe socket, 2 BNC sockets for connecting two selective electrodes—ISEs or pH, 2 banana sockets for connecting two reference electrodes, D-type 25 pin female connector for connection of printer, fuse, 3-pin mains chord and power ON/OFF switch. Reference electrodes are not required if the selective electrodes are combined type having internal reference electrodes.

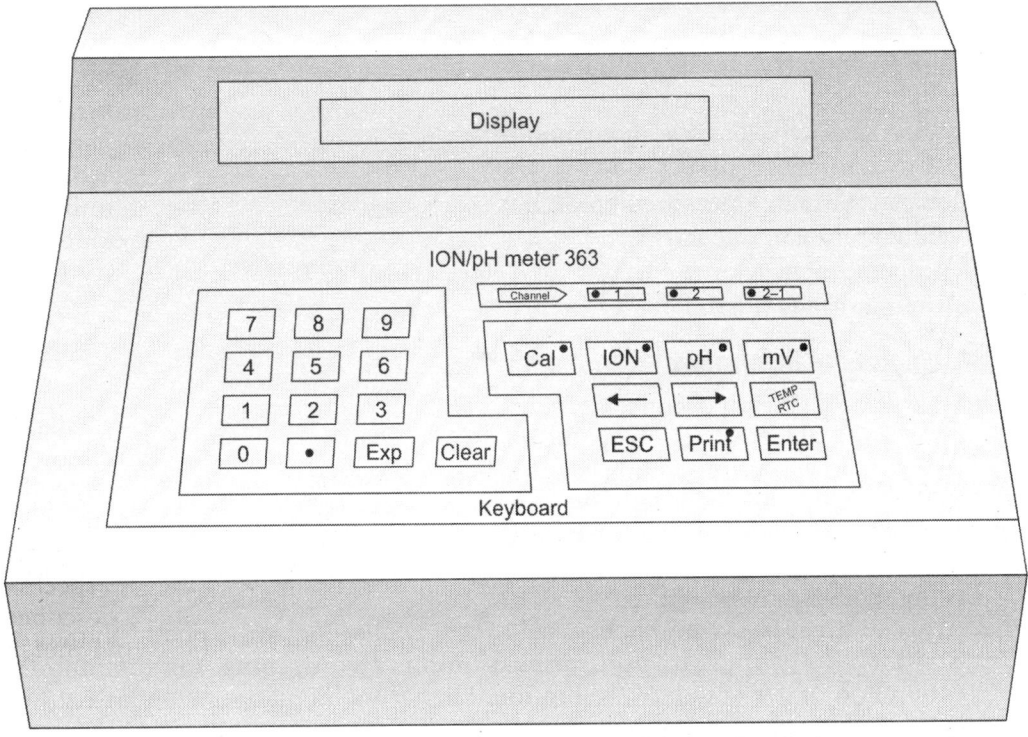

Fig. 6.1: Block diagram of ion/pH meter model 363 of systronics

b. Functions and Applications of Keys

The numeric keys are used for entering 0–9, dot (.) and exponential power. The uses of functional keys are as follows:

1. CLEAR—to clear wrong entry before ENTER key is pressed.
2. CAL—calibration mode—pH/ion and slope measurement.
3. ION—ion concentration mode (calibration and measurement).
4. pH—pH concentration mode (calibration and measurement).
5. mV—mV measuring mode.

6. TEMP/RTC—temperature measuring mode/real time clock.

7. →—moves the display screen forward.

8. ←—moves the display screen backward.

9. ESC—pressing it after error message or during measurement, takes the unit to the beginning of running routine/step. Pressing it at the end of measurement, takes the instrument to the initial starting stage.

10. PRINT—prints the result, if printer is attached.

11. ENTER—confirms the entry/selection.

c. Operating Modes

The operating modes are as follows:

1. Unit calibration (CAL)
   i. pH calibration mode
   ii. Ion concentration calibration mode
   iii. pH electrode slope mode
   iv. Ion selective electrode slope mode
2. pH measurement (pH)
3. Ion concentration measurement (ION)
4. Millivolt measurement (mV)
5. Temperature measurement/real time clock (TEMP/RTC)
6. Printing results (PRINT)

d. Starting of the Instrument

i. Connect pH/ISE to channel CH1 or CH2 BNC connector at back panel. If the concerned electrode is not combined type, connect the separate reference electrode to the corresponding banana socket. A second pH/ISE can be connected to the spare channel. The temperature probe can be connected to TEMP socket, if available. A message will be displayed.

ii. Select the mode, i.e. pH calibration and measurement or ion concentration calibration and measurement.

iii. pH calibration has two options, viz. 2 point calibration and 3 point calibration. Similarly ion concentration calibration has two options, viz. 2-standard calibration or 3–5 standard calibration. Higher the number of standards used, better is the accuracy of measurement.

iv. Different flowcharts are available in the instruction manual provided by the manufacturer. Follow the required flow chart for pH/ ion concentration calibration and measurement.

The pH of water samples can be determined by pH meter and water analyzer kit also.

6.1.2. Spectrophotometer

When a beam of monochromatic light falls on a substance, a fraction of light is absorbed and a fraction is transmitted. The absorbance and transmittance are proportional to the concentration of the substance. This property is used in spectrophotometer to measure the concentration of the ion in a solution, particularly when it is present in minor quantity. Different substances respond differently at different wavelengths (spectra). The block diagram of a spectrophotometer is shown in Fig. 6.2.

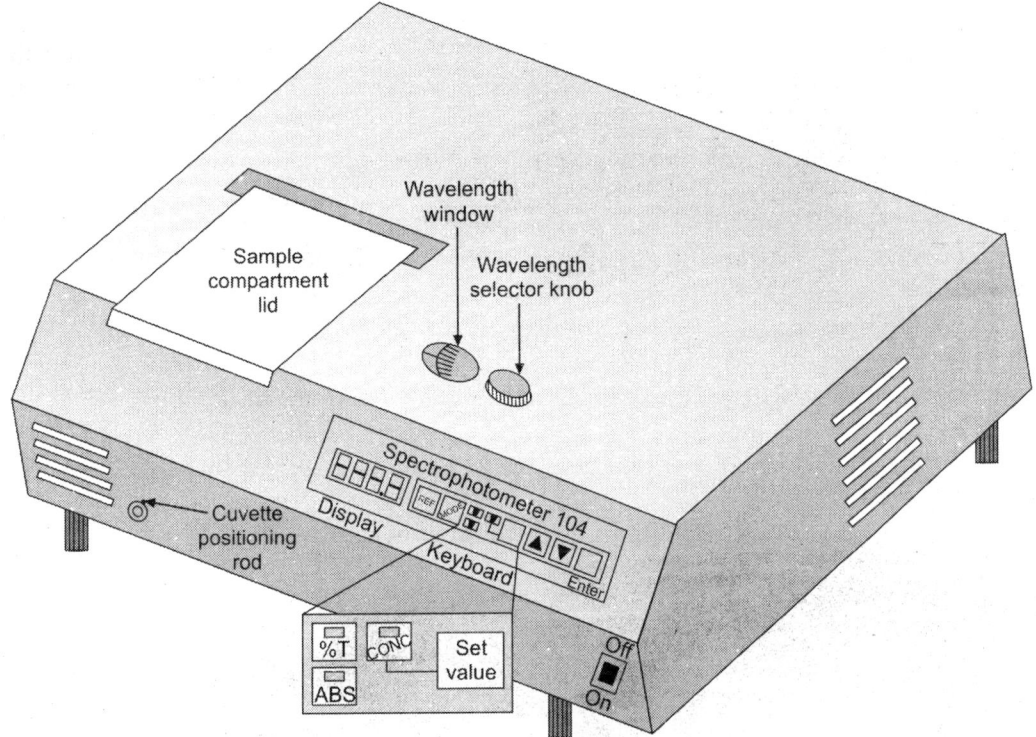

Fig. 6.2: Block diagram of spectrophotometer model 104 of systronics

a. Keys and their Functions

1. Display—it indicates 4-digit measurements as selected by MODE switch.
2. REF—used for setting references after putting reference solution.
3. Mode—selects transmittance (%T), absorbance (ABS) and concentration (CONC) modes. The selected mode is indicated by LCD.
4. Set vallue—it selects digits (1, 2, 3 and 4) to be changed for setting concentration.
5. ▲/▼—used to increase/decrease selected digit.
6. Enter—used to accept the value (4 and 5 above).
7. Wave length selector knob—sets desired wavelength.
8. Sample compartment lid—it is a hinged plate to close the cell compartment.
9. On/Off switch—connects/disconnects power supply.
10. Wavelength window

No. 2	340–390 nm
No. 4	390–630 nm
No. 6	Above 630 nm
Nos.1, 3, 5	Zero (Dark)

11. Filter wheel—this is located inside the sample compartment. It is used to select order cut-off filters manually.

12. Cuvettes positioning rod—spectrophotometer model 104 of Systronics made is provided with 10 mm length rectangular cuvettes. The cuvettes positioning rod is used for setting up of cuvettes so that the optically polished surface faces the photocell.

b. Operating Procedure

1. Insert the mains plug in 230 V AC outlet.
2. Switch on the instrument. After warming up for 20 minutes, the instrument will be ready.
3. Turn the wavelength selector knob to set the desired wavelength.
4. Select measuring mode [(transmittance (%T), absorbance (ABS) and concentration (CONC) modes)]
5. Press REF key ⇒ Display shows ZERO on data LCD.
6. Set filter position manually to 1 or 3 or 5 (Dark)
7. Press REF key ⇒ Display blinks CAL and finally shows rEF.
8. Depending upon the selected wavelength, choose the cut-off filter
9. Lift the lid and insert the Reference (Blank) solution cuvette (10 nm) in the cell holder. Close the lid slowly.
10. Press REF key ⇒ Display blinks CAL and finally shows 100 (%T) or 0.000 (ABS) depending upon the selection mode.
11. Put the sample solution in the cell holder. Display shows the measurement automatically.
12. In case of difficulty, follow the instruction manual provided by the manufacturer.

6.1.3. Water Analyzer Kit

The water analyzer kit is a micro-controller based instrument that can measure pH, dissolved oxygen, salinity, conductivity, temperature, total dissolved solids and turbidity in water sample. It is a portable instrument housed in a brief case, which can be carried to the field to measure the parameters at the sample collection site.

The block diagram of the front panel of water analyzer kit model—371 (systronics made) is shown in Fig. 6.3.

a. Panel Features

The panel features are as follows:
1. **Cell/battery compartment:** It contains battery, pH electrode, temperature sensor, glass conductivity cell with 1.0 and 0.1 cell constants, dissolved oxygen membranes and probe, colourimeter holder, big and small test tubes, glass filters and mains cord.
2. Display—2-line, 20-character LCD readout for interaction during setting up/programming.
3. Mains socket—inlet for 3-pin mains cord.
4. On/Off switch—to switch on/off the instrument.
5. Printer connector—provision for printer connection.
6. Fuse holder—55 mA mains fuse.
7. Battery charger on LED—indicates charging condition of battery.
8. Battery low on LCD—indicates battery charging is required.

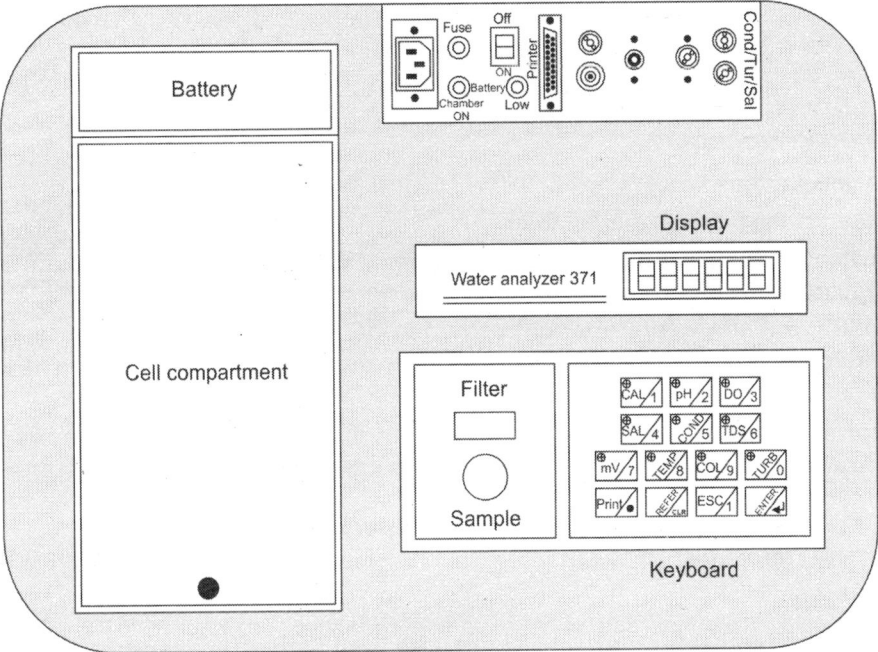

Fig. 6.3: Top view of water analyzer kit model 371 of systronics

9. pH/mV socket—inlet for pH or combined electrodes.

10 TEMP socket—inlet for temperature probe.

DO socket—inlet for dissolved oxygen probe.

COND/TDS/SAL—inlets for conductivity cell.

Keyboard—11 numeric keys [0–9 and dot (.)]

14 multi-function keys

1. CAL—selects calibration mode.
2. pH—selects pH mode for calibration/measurement.
3. DO—selects DO (dissolved oxygen) mode for calibration/ measurement.
4. SAL—selects SAL (salinity) mode for calibration/measurement.
5. COND—selects COND (conductivity) mode for calibration/ measurement.
6. TDS—selects TDS (total dissolved solids) mode for calibration/ measurement.
7. mV—selects mV (milli volt) mode.
8. TEMP—selects TEMP (temperature) mode.
9. COL—selects COL (colour) mode.
10. TURB—selects TURB (turbidity) mode.
11. PRINT—prints the data.
12. CLR/READ—clears wrong data/reads the last change in measured value.

13. ESC—skips the existing menu and goes to the main title or previous menu.
14. ENTER—confirms the entry.

b. Operating Procedure

1. Connect mains power cord to 230V AC outlet.
2. Connect the printer, if available.
3. Switch on the instrument. After warming up for 20 minutes, the instrument will be ready.
4. Follow the instruction manual provided by the manufacturer to measure the parameters for which calibration is not necessary or calibrate and measure other parameters.

The procedures of quantitative estimation of different parameters in surface and groundwater including environmental samples are given below.

6.2. ESTIMATION OF ALUMINIUM (Al) BY ERIOCHROME CYANINE R SPECTROPHOTOMETRIC METHOD

a. Apparatus

i. Spectrophotometer for use at 535 nm with light path of 1 cm or longer.
ii. Wash all the glassware with 1 + 1 warm HCl and rinse with aluminium free distilled water.

b. Reagents

i. **Stock aluminium solution:** Dissolve 8.791 g aluminium potassium sulphate, $AlK(SO_4)_2 \cdot 12H_2O$ in distilled water and dilute to 1 litre.
ii. **Standard aluminium solution:** Dilute 10 ml stock aluminium solution to 1000 ml with distilled water; 1 ml = 5 µg Al. Freshly prepared solution is to be used in each experiment.
iii. 0.02 N and 6 N sulphuric acid (H_2SO_4)
iv. **Ascorbic acid solution:** Dissolve 0.1 g ascorbic acid in water and make up to 100 ml in a volumetric flask. Freshly prepared solution is to be used in each experiment.
v. **Buffer reagent:** Dissolve 136 g sodium acetate $(NaC_2H_3O_2 \cdot 3H_2O)$ in water, add 40 ml 1 N acetic acid and dilute to 1 litre.
vi. **Stock indicator solution:** Prepare any one of the following solution.
   - Solochrome cyanine R-200 or Eriochrome cyanine: Dissolve 100 mg in water and dilute to 100 ml in a volumetric flask. This solution should have a pH of about 2.9.
   - Eriochrome cyanine R: Dissolve 300 mg in about 50 ml water. Adjust pH from about 9 to about 2.9 with 1+1 acetic acid (approximately 3 ml will be required). Dilute with distilled water to 100 ml.
   - Eriochrome cyanine R: Dissolve 150 mg in about 50 ml water. Adjust pH from about 9 to about 2.9 with 1 + 1 acetic acid (approximately 2 ml will be required). Dilute with distilled water to 100 ml.
vii. **Working indicator solution:** Dilute 10 ml of stock indicator solution to 100 ml in a volumetric flask with water. It can be used for about 6 months.

viii. Bromcresol green indicator, pH 4.5 solution: Dissolve 100 mg bromcresol green sodium salt in 100 ml distilled water.

ix. 0.01 M EDTA (disodium salt of ethylenediaminetetraacetic acid dihydrate): Dissolve 3.723 g in water and dilute to 1 litre.

x. 1 N and 0.1 N sodium hydroxide (NaOH).

c. Procedure

i. Prepare standards between 0 to 7 µg by taking 0, 1, 3, 5 and 7 ml standard aluminium solution in 50 ml volumetric flasks. Make the volume 25 ml in each case by adding distilled water.

ii. Add 1 ml 0.02 N H_2SO_4, 1 ml ascorbic acid solution and 10 ml buffer solution and 5 ml indicator reagent to each standard and mix thoroughly. Make up to 50 ml with distilled water and allow to stand for 5 to 10 minutes.

iii. Calibrate spectrophotometer to zero absorbance at 535 nm with the standard containing no aluminium. Read absorbance of standard solutions at 535 nm within 15 minutes of addition of indicator. Prepare a calibration curve between absorbance and aluminium concentration.

iv. Take 25 ml of water sample in a flask, add a few drops of bromcresol green and titrate with 0.02 N H_2SO_4 to yellowish end point. Record reading and throw away the sample.

vi. To two similar samples add 1 ml of acid excess than the amount used in the titration.

vii. To one sample add 1 ml disodium dihydrate salt of EDTA to complex any aluminium present. This will serve as blank. To both samples add 1 ml ascorbic acid, 10 ml buffer reagent, 5 ml working indicator, make up to 50 ml with distilled water and read absorbance at 535 nm.

d. Calculation

Read aluminium concentration in the sample against its absorbance value from the calibration curve. Deviation of the aluminium concentration between 10 and 50% occurs when this method is applied on samples that contain fluoride in the range of 0.4 and 1.5 mg/L.

6.3. ESTIMATION OF CALCIUM (Ca) BY EDTA TITRIMETRY

The amount of calcium present in water is determined by disodium dihydrate salt of EDTA titrimetry. Two similar methods are given below.

A. Method 1

a. Reagents

i. **1N NaOH solution:** Dissolve 1 g NaOH in 25 ml distilled water.

ii. **0.01M EDTA solution:** Dissolve 3.723 g disodium dihydrate salt of EDTA in 1000 ml of distilled water and add 2 NaOH pellets. The solution is to be standardized itl standard hard water.

iii. **Murexide (ammonium purpurate) indicator:** Mix 20 mg ammonium purpı at with 100 g solid NaCl. Grind to 40–50 mesh size.

b. Procedure

Take 10 ml of water sample; add 1 ml of NaOH and a little amount of murexide (1:5). Titrate with EDTA solution till pale violet coloured end point is reached.

c. Calculation

$$\text{Amount of Ca in mg/L} = \text{volume of EDTA in ml} \times 50$$

B. Method 2

a. Reagents

   i. **1 N sodium hydroxide (NaOH) solution:** Dissolve 40 g of NaOH in 1000 ml distilled water.

   ii. **Murexide (ammonium purpurate) indicator:** Mix 20 mg ammonium purpurate with 100 g solid NaCl. Grind to 40–50 mesh size.

   iii. **Standard EDTA solution, 0.01 M:** Dissolve 3.723 g disodium dihydrate salt of EDTA in distilled water, dilute to 1000 ml and store in polyethylene bottle, 1 ml EDTA solution = 400.8 µg Ca. Standardise EDTA against standard calcium solution periodically by the following method.

   Standard calcium solution: Take 1 g anhydrous $CaCO_3$ in 500 ml flask. Add 1 + 1 HCl in small amounts through a small funnel till all the $CaCO_3$ is completely dissolved. Add 200 ml distilled water and boil for a few minutes to expel CO_2. Cool and add a few drops of methyl red indicator and adjust to intermediate orange colour by adding 3 N NH_4OH or 1 + 1 HCl. Make the volume 1000 ml with distilled water. 1 ml EDTA solution = 400.8 µg Ca.

b. Procedure

   i. Take 50 ml water sample or a water sample diluted to 50 ml so that the calcium content is not more than 10 mg. Water of alkalinity greater than 300 mg/L should be neutralised with acid, boiled for 1 minute and cooled before titration.

   ii. Add 2 ml NaOH solution or a volume sufficient to produce a pH of 12 to 13. Add 0.1 to 0.2 g murexide indicator. Titrate with standard disodium dihydrate salt of EDTA solution, with continuous mixing, till the colour changes from pink to purple. Check the end point by adding 1 to 2 drops of titrant in excess to be sure that no further colour change occurs.

c. Calculation

$$\text{Ca in mg/L} = \frac{A \times B \times 400.8}{\text{Water sample in ml}}$$

where, A = titrant in ml

$$B = \frac{\text{ml of standard calcium solution taken for titration}}{\text{ml of EDTA titrant}}$$

Note: Although calcium concentration should be reported in mg/L, Calcium is sometimes also referred to as 'Calcium hardness'. Calcium hardness refers to the amount of calcium present in terms of $CaCO_3$ in mg/L = 2.5 × Ca in mg/L; (2.5 = 100 g $CaCO_3$/40 g Ca).

6.4. ESTIMATION OF CARBONATE AND BICARBONATE

The amount of carbonate and bicarbonate present in water sample can be determined both by titrimetry and computation methods.

6.4.1. Estimation of Carbonate and Bicarbonate by Titrimetry

Reagents required: 0.05 N H_2SO_4 (prepared by diluting 1.36 ml of conc. H_2SO_4 to 1000 ml with distilled water), phenolphthalein indicator and methyl orange indicator.

A. Procedure for Carbonate

Take 10 ml of water sample in a beaker and add 3 drops of phenolphthalein indicator. If the colour changes to pink, carbonate is present. Titrate against 0.05 N H_2SO_4 till colourless end point is reached.

Calculation

Carbonate in mg/L = Volume of acid in ml × 120

B. Procedure for Bicarbonate

Add 3 drops of methyl orange indicator that makes the water pink. Titrate against 0.05 N H_2SO_4 till pink colour changes to violet. If carbonate is absent, estimate bicarbonate directly.

Calculation

Bicarbonate in mg/L = Volume of acid in ml × 122

6.4.2. Estimation of Bicarbonate from pH and Alkalinity

a. Procedure

Determine values of pH, phenolphthalein alkalinity (P) and total alkalinity (T).

b. Calculation

   i. In case total dissolved solids less than 500 mg/L

$$HCO_3^- \text{ as } CaCO_3 \text{ in mg/L} = \frac{T - 5.0 \times 10^{(pH-10)}}{1 + 0.94 \times 10^{(pH-10)}}$$

   ii. In case total dissolved solids more than 500 mg/L.

Calculate bicarbonate from phenolphthalein alkalinity (P) and total alkalinity (T) both in $CaCO_3$ in mg/L as given below.

Alkalinity	Bicarbonate in mg/L	Alkalinity	Bicarbonate in mg/L
P = 0	T	P > ½ T	0
P < ½ T	T-2P	P = T	0
P = ½ T	0		

Convert to desired units: HCO_3^- (mg/L) = HCO_3^- ($CaCO_3$ in mg/L) × 1.22

6.4.3. Estimation of Carbonate from pH and Alkalinity

i. In case total dissolved solids less than 500 mg/L

$$HCO_3^- \text{ as } CaCO_3 \text{ in mg/L} = \frac{T - 5.0 \times 10^{(pH-10)}}{1 + 0.94 \times 10^{(pH-10)}}$$

$$HCO_3^{2-} \text{ as } CaCO_3 \text{ in mg/L}) = 0.94 \times HCO_3 \times 10^{(pH-10)}$$

where, T = total alkalinity as $CaCO_3$ in mg/L, HCO_3^- = bicarbonate

ii. In case total dissolved solids more than 500 mg/L

Calculate carbonate from phenolphtalein alkalinity (P) and total alkalinity (T) both in $CaCO_3$ mg/L as follows:

Alkalinity	Carbonate in mg/L	Alkalinity	Carbonate in mg/L
P = 0	0	P > ½ T	2 (T − P)
P < ½ T	2P	P = T	0
P = ½ T	2P		

Convert to desired units: CO_3 (mg/L) = CO_3 ($CaCO_3$ in mg/L) × 0.6
(Because 1 mole = 100 g $CaCO_3$ = 60 g CO_3)

6.5. ESTIMATION OF CHLORIDE

Three methods, one for water without interfering substances, second, for water sample with or without interfering substances and may be coloured and third, when presence of chlorine in distilled water is suspected are described below.

6.5.1. Water Sample without Interfering Substances
a. Reagents

i. **Standard AgNO₃ solution:** Dissolve 4.8 g of $AgNO_3$ in 950 ml of distilled water

ii. **Standard K₂CrO₄ solution:** Dissolve 5 g of K_2CrO_4 in 100 ml of distilled water.

b. Procedure

Take 10 ml of water sample in a beaker; add 2 drops of K_2CrO_4 solution that serves as indicator. Titrate against standard $AgNO_3$ solution till the sample turns brick red that persists for 10–15 seconds.

Calculation

$$\text{Chloride in mg/L} = \text{Volume of } AgNO_3 \text{ solution in ml} \times 100$$

6.5.2. Water sample with or without Interfering Substances and may be Coloured
a. Reagents

i. **Standard chloride solution:** Dissolve 0.16486 g of NaCl (AR grade) in 1000 ml distilled water (1 ml = 100 μg Cl)

ii. **Standard AgNO₃ solution:** Dissolve 1.6987 g of AR grade $AgNO_3$ in 1000 ml of distilled water. The solution is to be kept in a brown bottle. If brown bottle is not available, white reagent bottle wrapped by a black paper should be used. The $AgNO_3$ solution is to be standardized by the procedure mentioned below.

iii. **Potassium chromate solution:** Dissolve 5 g of K_2CrO_4 in 100 ml distilled water.

iv. **Aluminium hydroxide suspension solution:** Dissolve 125 g of aluminium potassium sulphate $[AlK(SO_4)_2 \cdot 12H_2O]$ or aluminium ammonium sulphate $[AlNH_4(SO_4)_2 \cdot 12H_2O]$ in water. Warm and add 55 ml conc. NH_3 solution with constant stirring. Allow the solution to stand for 1 hour. Decant the supernatant floating on the water surface. Repeat the process at least three times to remove all the chloride ions. Make the volume of the $Al(OH)_3$ 1000 ml by adding distilled water.

v. 30% H_2O_2 solution.

vi. 1 N H_2SO_4

vii. 1 N NaOH—40 gram of NaOH in 1 liter distilled water

b. Standardisation of AgNO₃ solution

i. Take about 50 ml (V_1) of standard NaCl solution in a 250 ml beaker.
ii. Add distilled water to make the volume 100 ml.
iii. Adjust the pH to about 7 by adding H_2SO_4 or NaOH.
iv. Add 1 ml K_2CrO_4 indicator solution.
v. Titrate against prepared $AgNO_3$ solution till dirty orange colour appears. Say the volume of $AgNO_3$ solution consumed is V_2.

Calculation

$$\text{Normality of AgNO}_3 \text{ solution} = \frac{V_1}{V_2} \times \text{Nomality of NaCl solution}$$

6.5.3. Procedure for Colourless water Sample without any Interfering Substances

i. Take about 50 ml of water sample in a 250 ml beaker.
ii. Make the pH about 7 by adding H_2SO_4 or NaOH.
iii. Titrate against 0.1 N $AgNO_3$ (16.987 g of $AgNO_3$ in 1 liter distilled water) solution till dirty orange colour appears.

6.5.4. Procedure for Coloured Water Sample with Interfering Substances

i. Take about 50 ml water in a 250 ml beaker.
ii. Add 3–4 ml of $Al(OH)_3$ suspension, stir thoroughly and allow to stand for about 30 minutes. Filter the solution and determine the amount of chlorides in the filtrate.
iii. Titrate against 0.1 N $AgNO_3$ solution till dirty orange colour appears.

If you are sure that sulphide, sulphite and thiosulphate are present in the water sample, proceed as mentioned below.

i. Add 1 ml of H_2O_2 solution, stir for 2 minutes and make the pH nearly 7 by adding H_2SO_4 or NaOH.
ii. Add 1 ml of K_2CrO_4 solution and titrate against 0.1 N $AgNO_3$ solution till dirty orange colour appears.

Calculation

$$Cl^- \text{ in mg/L} = \frac{(35450 \times \text{volume of AgNO}_3 \times \text{Normality of AgNO}_3)}{\text{Volume of water sample}}$$

6.5.5. Distilled Water is Suspected to Contain Chlorine

a. Reagents

i. **Potassium chromate indicator solution:** Dissolve 50 g K_2CrO_4 in small amount of distilled water. Add $AgNO_3$ solution till red precipitate is formed. Allow to stand for 12 hours, filter and dilute to 1 litre with distilled water.

ii. **Standard sodium chloride solution, 0.0141N:** Dissolve 824.0 mg NaCl (dried at 140°C) in distilled water and dilute to 1000 ml; 1 ml = 500 µg Cl^-.

iii. **Standard silver nitrate titrant (0.0141N):** Dissolve 2.395 g $AgNO_3$ in distilled water and dilute to 1000 ml; 1 ml = 500 µg Cl^-. Store the solution in a brown bottle.

Standardise against 10 ml standard NaCl diluted to 100 ml

$$N = 0.0141 \times \frac{\text{Standard NaCl in ml}}{A - B}$$

where,

N = normality of $AgNO_3$
A = $AgNO_3$ titrant in ml against standard NaCl
B = $AgNO_3$ titrant in ml against blank (distilled water)

iv. **Aluminium hydroxide suspension:** Dissolve 125 g aluminium potassium sulphate [$AlK(SO_4)_2 \cdot 12H_2O$] or aluminium ammonium sulphate [$AlNH_4(SO_4)_2 \cdot 12H_2O$] in 1 litre distilled water. Warm to 60°C and add 55 ml conc. ammonium hydroxide slowly with stirring. Allow to stand for about 1 hour and transfer to a large bottle. Wash the precipitate twice or till free of chloride, by successive addition of distilled water, settling and decanting.

b. Procedure

i. Use a 100 ml water sample or a suitable amount of water sample diluted to 100 ml. If the sample is coloured or turbid, add 3 ml $Al(OH)_3$ suspension, mix thoroughly, allow to settle and filter.

ii. Add 1 ml K_2CrO_4 indicator solution, titrate with $AgNO_3$ to a pinkish yellow end point. Say volume of $AgNO_3$ used is A.

iii. Repeat the titration with 0.2 to 0.3 ml blank (distilled water). Say volume of $AgNO_3$ used is B.

Calculation

$$Cl^- \text{ in mg/L} = \frac{(A - B) \times N \times 35450}{\text{ml of sample}}$$

where,

A = ml of $AgNO_3$ solution used for water sample,
B = ml of $AgNO_3$ solution used for blank (distilled water),
N = normality of $AgNO_3$ solution

6.6. ESTIMATION OF HEXAVALENT CHROMIUM

a. Reagents

i. **Chromium stock solution:** Dissolve 0.2828 gm $K_2Cr_2O_7$ in distilled water and dilute to 1000 ml with distilled water. (1 ml = 100 µg Cr)

ii. **Standard chromium solution:** Dilute 1 ml of 100 µg Cr solution to 100 ml volumetric flask (1 ml = 1 µg Cr).

iii. **1,5-diphenylcarbazide solution:** Dissolve 0.5 g 1,5-diphenylcarbazide in 100 ml acetone.

iv. **0.2 N sulphuric acid:** Dilute 5.6 ml conc. H_2SO_4 to 1000 ml.

v. **Phosphoric acid:** Concentrated and 70% solution.

vi. **Sodium azide solution:** Dissolve 0.5 g NaN_3 in 100 ml distilled water.

b. Preparation of standard graph

i. Take a suitable amount of aliquot from standard chromium solution in the range of 2–10 µg Cr in 100 ml volumetric flasks.

ii. Add 2 ml (70%) phosphoric acid solution in each volumetric flask and mix.

iii. Add 2 ml diphenylcarbazide solution and mix.

iv. Make up the volume to 100 ml with 0.2N H_2SO_4 and mix thoroughly.

v. Allow to stand for 5–10 minutes to develop a colour.

vi. Measure absorbance at 540 nm.

vii. Plot a graph of concentrations against absorbances.

c. Measurement of hexavalent chromium

6.6.1. Procedure in Absence of Oxidizing or Reducing Substances

i. If any suspended particles are present, filter the water sample through 0.45 µm filter paper.

ii. Take suitable amount of filtered sample in a 100 ml volumetric flask.

iii. Add 2 ml (70%) phosphoric acid solution and mix.

iv. Add 2 ml diphenylcarbazide solution and mix.

v. Dilute to 100 ml with 0.2N H_2SO_4 and mix thoroughly.

vi. Allow to stand for 5–10 minutes to develop a colour.

vii. Measure absorbance at 540 nm.

viii. Find out the concentration of chromium from a standard graph.

6.6.2. Procedure in the Presence of Oxidizing or Reducing Substances

a. Reagents

i. **Sodium hypochlorite solution:** Dilute 70 ml sodium hypochlorite solution (NaOCl) to 1000 ml with distilled water.

ii. Potassium iodide starch test paper.

iii. **Phosphoric acid solution:** Dilute 700 ml phosphoric acid to 1000 ml with distilled water.

iv. **Sodium chloride:** AR grade NaCl.

v. **1,5-diphenylcarbazide solution:** Dissolve 0.1 g 1,5-diphenylcarbazide in acetone and dilute to 100 ml.

b. Procedure

i. Take suitable amount of the filtered sample in 100 ml volumetric flask.

ii. Add 1 ml sodium hypochlorite solution and mix.

iii. Check the excess of chlorine using potassium iodide starch paper. If there is no excess of chlorine, add more sodium hypochlorite solution until chlorine becomes slightly excess.

iv. Add 2 ml phosphoric acid solution and mix.

v. Add 10 g NaCl and mix thoroughly till the NaCl dissolves completely.

vi. Pass air through the solution at the rate of approximately 40 L/h for about 40 minutes in fuming chamber.

vii. Add 2 ml diphenylcarbazide solution, mix thoroughly and dilute to 100 ml.

viii. Measure the absorbance after 5–10 minutes at 540 nm.

c. Calculation

$$Cr^{6+} \text{ in } \mu g/L = \frac{Cr \text{ in } \mu g \text{ in water sample}}{\text{Amount of water sample analysed in ml}} \times 1000$$

6.7. DETERMINATION OF COLOUR OF WATER SAMPLE BY UV-SPECTROPHOTOMETRIC METHOD

Colour may be due to presence of soluble ions like iron and/or manganese or any other material in water. Generally colour is determined by comparing with that of known standards of colour units.

a. Reagents

i. Potassium chloroplatinate solution: Dissolve 0.2492 g potassium chloroplatinate (K_2PtCl_6) and 0.2 g cobaltous chloride ($CoCl_2 \cdot 6H_2O$) in 10 ml conc. HCl and dilute to 100 ml with distilled water.

$$(1 \text{ ml} = 1 \text{ mg Pt} = 1000 \text{ } \mu g \text{ Pt} = 1000 \text{ Hazen})$$

b. Preparation of standard graph

Take suitable quantity of standard potassium chloroplatinate solutions ranging from 0–70 Hazen units in 100 ml volumetric flasks and dilute to 100 ml. Measure absorbance/transmittances at 288 nm. Plot standard graph between absorbance and Hazen units.

c. Measurement of colour of sample solution

Measure the absorbance of water sample at 288 nm. If absorbance exceeds 70 Hazen, sample solution is to be diluted so that absorbance comes in the range of 70 Hazen units. Turbid water samples are to be centrifused to remove turbidity before measuring absorbance.

6.8. ESTIMATION OF CYANIDE

Cyanide can be estimated in water sample both by ion selective electrode and colourimetric methods.

6.8.1. Estimation of Cyanide by Ion Selective Electrode Method

a. Reagents

i. **Standard cyanide solution:** Dissolve 1.88 g AR grade NaCN in water and dilute to 1000 ml with distilled water in polythene volumetric flask. (1 ml = 1000 mg/L = 1000 ppm CN).

ii. **Ionic strength buffer** (0.2 M NaOH): Dissolve 8 g NaOH in water with constant stirring. Cool and dilute to 1000 ml in polythene volumetric flask.

b. Preparation of standard graph

i. Take suitable volume of standard stock solution of cyanide and prepare series of 4–5 solutions of 0.2 ppm, 0.4 ppm, 0.6 ppm, 0.8 ppm and 1.0 ppm CN solution in 100 ml volumetric flasks.

ii. Take 25 ml aliquot of above solutions in 100 ml polythene beaker.

iii. Add equal volume (25 ml) of 0.2 M NaOH to each beaker.

iv. Place one magnetic stirrer bar in the solution.

v. Place a beaker containing solution 0.2 ppm on a magnetic stirrer.

vi. Immerse the cyanide and reference electrodes in the solution.

vii. Switch on the ion meter and magnetic stirrer. Allow the meter reading to stabilise. Note the reading.

viii. Repeat the process for other solutions of higher concentrations as prepared above.

ix. Plot a standard graph between concentrations and readings shown by the meter for each concentrations.

c. Measurement of cyanide concentration in water sample

i. Take 25 ml or 50 ml of water sample in 100 ml polythene beaker.

ii. Add equal volume of 0.2 M NaOH buffer (25 ml or 50 ml).

iii. Place one magnetic stirrer bar in the solution.

iv. Place the beaker containing solution on a magnetic stirrer.

v. Immerse the cyanide and reference electrodes in the solution.

vi. Switch on the ion meter and magnetic stirrer. Allow the meter reading to stabilise. Note the reading.

vii. Find out the concentration of cyanide from the standard graph.

d. Calculation

$$CN \text{ in } \mu g/L = \frac{CN \text{ in mg in water sample}}{\text{Volume of water sample}} \times 1000$$

6.8.2. Estimation of Cyanide by Colorimetric Method

a. Reagents

i. **Chloramine-T:** Dissolve 1 g chloramine-T powder in water and dilute to 100 ml.

ii. **Standard cyanide solution:** Dissolve 1.88 g AR grade NaCN in water and dilute to 1000 ml in a polythene volumetric flask with distilled water (1 ml = 1000 mg/L = 1000 ppm CN).

iii. Alternatively dissolve 0.188 g AR grade NaCN in water and dilute to 1000 ml in a polythene volumetric flask with distilled water (1 ml = 100 ppm CN).

iv. Dilute above CN solution so that 1 ml = 1 μg

v. **Pyridine barbituric acid solution:** Dissolve 6 g barbituric acid in minimum quantity of water, add 30 ml pyridine and 6 ml concentrated HCl and mix thoroughly. Cool and dilute to 100 ml in polythene volumetric flask.

vi. **Sodium dihydrogen phosphate (1M):** Dissolve 138 g NaH_2PO_4 in water and dilute to 1000 ml in a polythene volumetric flask.

vii. **Sodium hydroxide solution:** Dissolve 1.6 g NaOH in water, cool and dilute to 1000 ml in polythene volumetric flask.

b. Preparation of standard graph

i. Take appropriate volume of the standard stock solution to prepare series of standards ranging from 2 μg to 10 μg in 100 ml polythene volumetric flasks.

ii. Add 4 ml sodium dihydrogen phosphate buffer, 2 ml chloramines-T solution and 5 ml pyridine-barbituric acid solution in succession to each flask, mix thoroughly each time and dilute to 100 ml.

iii. Measure the transmittances/absorbance at 578 nm using spectrophotometer of 1 cm cell path.

iv. Plot a graph between cyanide concentrations and absorbances.

c. Measurement of cyanide in water sample

i. Take suitable aliquot of sample in 100 ml in polythene volumetric flask add 4 ml phosphate buffer, 2 ml chloramines-T solution, 5 ml pyridine-barbituric acid solution in succession and mix thoroughly. Make up the volume to 100 ml with distilled water.

ii. Measure the absorbance at 578 nm using spectrophotometer.

iii. Calculate CN concentration from the standard graph.

d. Calculation

$$\text{CN in mg/L} = \frac{\text{CN in mg in water sample}}{\text{Volume of water sample}} \times 1000$$

6.9. ESTIMATION OF ELECTRICAL CONDUCTIVITY (EC) BY CONDUCTIVITY METER

a. Apparatus

i. Conductivity meter capable of measuring conductivity with an error not exceeding 1 % or 0.1 mS/m which ever is greater.

ii. Platinum electrode type conductivity cell: New cells not coated and old cell giving erratic readings are to be platinised. Clean the cell with chromic-sulphuric acid cleaning mixture. Prepare platinising solution by dissolving 1 g chloroplatinic acid ($H_2PtCl_6·6H_2O$) and 12 mg lead acetate in 100 ml distilled water. Immerse electrodes in this solution and connect both to the negative terminal of a 1.5 V dry cell battery (in some meters this source is inbuilt). Connect the positive terminal to a platinum wire and dip the wire into the solution. Continue electrolysis until both cell electrodes are coated with platinum black. In many instances, ready to use conductivity cells are provided by the manufacturer.

b. Reagent

i. Distilled water: Use distilled water boiled shortly before use to minimise CO_2 content. Electrical conductivity must be less than 0.1 µmho/cm.

ii. Standard potassium chloride solution (KCl), 0.01M, conductivity 1412 µmho/cm at 25°C: Dissolve 745.6 mg anhydrous KCl (dried 1 hour at 180°C) in distilled water and dilute to 1000 ml. This reference solution is suitable when the cell has a constant between 1 and 2 per cm.

c. Procedure

i. Rinse the conductivity cell with at least three portions of 0.01 M KCl solution. Measure resistance of a fourth portion and note the temperature.

ii. If the instrument indicates conductivity directly and has internal temperature compensation facility, after rinsing adjust temperature compensation dial to 0.0191/°C with the probe in standard KCl solution. Adjust meter to read 1412 µmho/cm.

iii. Compute the cell constant, K_c according to the formula:

$$K_c = \frac{1412}{C_{KCl}} \times [0.0191(T - 25) + 1]$$

where,

K_c = The cell constant in cm^{-1}

C_{KCl} = Measured conductance in µ mho

T = Observed temperature of standard KCl solution in °C

iv. Rinse the cell by water sample 2–3 times. The level of sample must be above the vent holes in the cell and there should be no air bubbles inside the cell. Adjust the temperature of sample to about 25°C. If the temperature is beyond the range of 20–30°C, error creeps in which proportional to the deviation from 25°C.

v. Read conductivity of water sample and note temperature to nearest 0.1°C.

vi. Rinse the cell thoroughly in distilled water after measurement and keep it in distilled water when not in use.

d. Calculation

When conductivity of water sample is measured with instruments having temperature compensation facility, the reading is automatically corrected to 25°C. If the instrument lacks internal temperature compensation facility, necessary correction is to be made. Electrical conductivity (µmho/cm) at 25°C is:

$$\frac{C_M \times K_c}{[0.0191(T - 25) + 1]}$$

where,

C_M = Measured conductance of the sample in µmho

K_c = The cell constant in cm^{-1} and

T = Observed temperature of sample in °C

Table 6.1: Value of [0.0191 × (T–25) +1] for temperature correction

T (°C)	0.0	0.1	0.2	0.3	0.4	0.5	0.6	0.7	0.8	0.9
15	0.810	0.812	0.814	0.816	0.818	0.820	0.821	0.823	0.825	0.827
16	0.829	0.831	0.833	0.835	0.837	0.839	0.840	0.842	0.844	0.846
17	0.848	0.850	0.852	0.854	0.856	0.858	0.859	0.861	0.863	0.865
18	0.867	0.869	0.871	0.873	0.875	0.877	0.878	0.880	0.882	0.884
19	0.886	0.888	0.890	0.892	0.894	0.896	0.897	0.899	0.901	0.903
20	0.905	0.907	0.909	0.911	0.913	0.915	0.916	0.918	0.920	0.922
21	0.924	0.926	0.928	0.930	0.932	0.934	0.935	0.934	0.939	0.941
22	0.943	0.945	0.947	0.949	0.951	0.953	0.954	0.956	0.958	0.960
23	0.962	0.964	0.966	0.968	0.70	0.972	0.973	0.975	0.977	0.979
24	0.981	0.983	0.985	0.987	0.989	0.991	0.992	0.994	0.996	0.998
25	1.000	1.002	1.004	1.006	1.008	1.010	1.011	1.013	1.015	1.017
26	1.019	1.021	1.023	1.025	1.027	1.029	1.030	1.032	1.034	1.036
27	1.038	1.040	1.042	1.044	1.046	1.048	1.049	1.051	1.053	10.55
28	1.057	1.059	1.061	1.063	1.065	1.067	1.069	1.070	1.072	1.074
29	1.076	1.078	1.080	1.082	1.084	1.086	1.087	1.089	1.091	1.093
30	1.095	1.097	1.099	1.101	1.103	1.105	1.106	1.108	1.110	1.112
31	1.114	1.116	1.118	1.120	1.122	1.124	1.125	1.127	1.129	1.131
32	1.133	1.135	1.137	1.139	1.141	1.143	1.144	1.146	1 .448	1.150
33	1.152	1.154	1.156	1.158	1.160	1.162	1.163	1.165	1.167	1.169
34	1.171	1.173	1.175	1.177	1.179	1.181	1.182	1.184	1.186	1.188
35	1.190	1.194	1.194	1.196	1.198	1.200	1.201	1.203	1 .205	1.207

The value of temperature correction [0.0191 × (T–25) + 1] can be read from Table 6.1 Report conductivity preferably in µmho/cm. Note that 1 S (Siemens) = 1 mho.

Electrical conductivity is a function of total dissolved solids (TDS = 0.64 × EC) and can be measured by water analyser.

6.10. ESTIMATION OF FLUORIDE

The fluoride content of the water can be estimated by ion selective electrode, SPADNS spectrophotometric and colorimetric methods.

6.10.1. Estimation of Fluoride (F) by Ion Selective Electrode Method

a. Apparatus

   i. Ion meter
  ii. Fluoride and reference electrodes
 iii. Magnetic stirrer with TFE coated stirring bar

b. Reagents

   i. **Stock fluoride solution:** Dissolve 221 mg anhydrous sodium fluoride (NaF) in distilled water and dilute to 1000 ml. 1 ml = 100 µg F⁻.

ii. **Standard fluoride solution:** Dilute 100 ml stock fluoride solution to 1000 ml with distilled water. 1 ml = 10 μg F⁻.

iii. **Fluoride buffer:** Take about 500 ml distilled water in a 1 litre beaker; add 57 ml glacial acetic acid, 58 g NaCl and 4 g 1,2 cyclohexylenediaminetetraacetic acid. Keep the beaker in a cool bath and slowly add about 125 ml 6 N NaOH with stirring until the pH becomes 5.3 to 5.5. Transfer the solution to the 1 litre volumetric flask and add distilled water to make the volume 1 litre.

c. Procedure

i. Prepare a series of working standards by diluting 5, 10 and 20 ml of standard solution to 100 ml, corresponding to 0.5, 1.0 and 2 mg F⁻/L, respectively.

ii. Take 10 to 25 ml standards and water sample in 100 ml beakers. Bring the samples and the standards to the room temperature and add an equal volume of buffer to each beaker. The total volume should be sufficient to immerse the electrode and allow the use of the stirrer.

iii. Follow manufacturer's instructions to set up and calibrate the ion meter using standards in the prescribed range. In many cases standards diluted with the buffer are supplied with the ion meter by the manufacturer. Do not stir before immersing the electrodes to avoid entrapment of air bubbles.

iv. If a direct reading instrument is not used, plot potential measurement of fluoride standards on arithmetic scale and fluoride concentration on logarithmic scale on a semilogarithmic graph paper.

d. Calculation

Read fluoride concentration in the water sample from the calibration curve or directly from the meter.

6.10.2. Estimation of Fluoride by SPADNS Spectrophotometric Method

a. Apparatus

i. **Distillation apparatus:** 1 litre round bottom long neck, borosilicate glass boiling flask, connecting tube and an efficient condenser, with thermometer adapter and a thermometer reading up to 200°C (Fig. 6.4).

ii. **Spectrophotometer for use at 570 nm:** It must provide a light path of at least 1 cm or a spectrophotometer with a greenish yellow filter (550 to 580 nm).

b. Reagents

i. **Conc. sulphuric acid (H_2SO_4)**

ii. **Silver sulphate (Ag_2SO_4) crystals**

iii. **Stock fluoride solution:** Dissolve 221 mg anhydrous sodium fluoride (NaF) in distilled water and dilute to 1000 ml. 1 ml = 100 μg F⁻.

iv. **Standard fluoride solution:** Dilute 100 ml stock fluoride solution to 1000 ml with distilled water. 1 ml = 10 μg F⁻.

v. **SPADNS solution:** Dissolve 958 mg SPADNS [sodium 2-(parasulphophenylazo) 1,8-dihydroxy-3,6-naphthalenedisulphonate] in distilled water and dilute to 500 ml. Store in a black glass bottle so that it can be protected from light and will remain stable for 1 year.

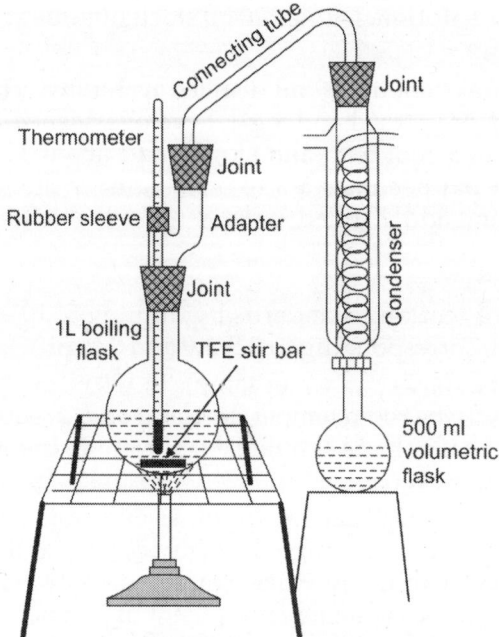

Fig. 6.4: Fluoride distillation apparatus

vi. **Zirconyl-acid reagent:** Dissolve 133 mg zirconyl chloride octahydrate ($ZrOCl_2 \cdot 8H_2O$) in about 25 ml distilled water, add 350 ml conc. HCl and dilute to 500 ml.

vii. **Mixed acid zirconyl-SPADNS reagent:** Mix equal volumes of SPADNS solution and zirconyl-acid reagent.

viii. **Reference solution:** Add 10 ml SPADNS solution to 100 ml distilled water. Dilute 7 ml conc. HCl to 10 ml with distilled water and add to SPADNS solution. Set the spectrometer to zero with this solution.

ix. **Sodium arsenite solution:** Dissolve 5 g $NaAsO_2$ and dilute to 1 litre with distilled water (sodium arsenite solution is toxic, so it is to be handled carefully).

c. Procedure

i. **Distillation:** Distillation is necessary for samples containing high concentration of dissolved solids (proceed to step iv. If distillation is not required). Take 400 ml distilled water in the distillation flask, with operating magnetic stirrer, add 200 ml conc. H_2SO_4 and a few glass beads. Connect the apparatus as shown in the Fig.6.4 and heat to 180°C. Prevent overheating by stopping heating when temperature reaches 178°C. Discard the distillate.

ii. Cool the acid mixture remaining in the flask to 80°C and add 300 ml water sample. With stirrer operating, distil until the temperature reaches 180°C (again stop heating at 178 °C to prevent overheating), turn off heat; retain the distillate for analysis.

iii. Add $AgSO_4$ to the distilling flask at the rate of 5 mg/mg Cl⁻ to avoid Cl⁻ interference. H_2SO_4 solution in the flask can be used continually until impurity from samples collect to such an extent that recovery is hindered.

iv. **Standard curve preparation:** Take the following volumes of standard fluoride solution and dilute to 50 ml with distilled water and note down the temperature:

Standard F⁻ solution in ml	0	0.1	0.2	0.5	1.0	2.0	3.0	5.0	7.0
F⁻ in µg	0	1	2	5	10	20	30	50	70

v. Pipette 10 ml of mixed acid—zirconyl-SPADNS reagent to each standard and mix well. Set spectrophotometer to zero absorbance with the reference solution and note absorbance readings of standards at 570 nm. Plot a curve of mg F⁻ versus absorbance. Prepare a new standard curve whenever a fresh reagent or a different standard temperature is used.

vi. **Sample pre-treatment:** If the sample contains residual chlorine remove it by adding 1 drop (0.05 ml) $NaAsO_2$ solution/ 0.1 mg residual chlorine.

vii. **Colour development:** Take a 50 ml sample or a portion diluted to 50 ml with distilled water. Adjust sample temperature to that of the standard curve. Set reference point of spectrophotometer as above. Add 10 ml acid–zirconyl-SPADNS reagent, mix well and read the absorbance. If the absorbance falls beyond the range of the standard curve repeat using a diluted sample.

d. Calculation

$$F^- \text{ in mg/L} = \frac{A}{B \times R}$$

where,

A = F⁻ reading from the standard curve in µg

B = Volume of diluted or undiluted sample taken for colour development in ml

R = Ratio of the volume of sample taken for dilution to the final volume after dilution, when sample is diluted

6.10.3. Estimation of Fluoride in Water Sample by Colorimetric Method

a. Reagents

i. **Standard fluoride solution:** Dissolve 0.221 g of NaF in distilled water and dilute to 1000 ml. (1 ml of solution = 100 ppm F⁻)

ii. **SPADNS solution:** Dissolve 0.1916 g of SPADNS reagent in distilled water and dilute to 100 ml.

iii. **Zirconyl-acid reagent:** Dissolve 0.133 g of zirconyloxychlorideoctahydrate $(ZrOCl_2 \cdot 8H_2O)$ in distilled water, add 350 ml conc. HCl and dilute to 500 ml by adding distilled water.

iv. **SPADNS-Zirconyl-acid solution:** Mix equal volumes of SPADNS solution and Zirconyl–acid reagent.

v. **Reference solution:** Dilute 10 ml SPADNS solution to 100 ml and 7 ml conc. HCl to 10 ml. Mix these two solutions thoroughly. This solution mixture is to be used for setting zero of the spectrophotometer.

vi. **Arsenite solution:** Dissolve 0.5 g of sodium arsenite $(NaAsO_3)$ in distilled water and make the volume 100 ml with distilled water.

b. Preparation of standard graph

　　i. Take 0.2 ppm, 0.4 ppm, 0.6 ppm, 0.8 ppm, 1.0 ppm, 1.2 ppm and 1.4 ppm fluoride solutions in 100 ml flasks.

　　ii. Add 10 ml SPADNS-Zirconyl–acid solution to each of the fluoride solutions.

　　iii. Measure the absorbances/transmittances of each solution at 570 nm.

　　iv. Plot a graph of concentration versus absorbance/transmittance.

c. Estimation of fluoride content

　　i. Confirm the presence of residual chlorine in water sample. In case residual chlorine is present, remove it by adding a drop of arsenite solution.

　　ii. Take 50 ml of water sample in 100 ml flask, add 10 ml SPADNS-Zirconyl–acid solution and mix thoroughly.

　　iii. Determine the absorbance/transmittance at 570 nm.

　　iv. Find out the fluoride concentration from the standard graph.

6.11. ESTIMATION OF IRON (Fe) BY PHENANTHROLINE SPECTROPHOTOMETRIC METHOD

a. Apparatus

　　i. Spectrophotometer for use at 510 nm with a light path of 1 cm or longer.

　　ii. **Separatory funnels:** 125 ml ground glass or TFE stopcocks and stoppers

b. Reagents

　　i. Conc. HCl with less than 0.00005% iron.

　　ii. **Hydroxylamine solution:** Dissolve 10 g $NH_2OH \cdot HCl$ in 100 ml distilled water.

　　iii. **Ammonium acetate buffer solution:** Dissolve 250 g $NH_4C_2H_3O_2$ in 150 ml water and add 700 ml glacial acetic acid. Since ammonium acetate may contain a significant amount of iron, prepare new reference standards with each buffer preparation.

　　iv. **Sodium acetate solution:** Dissolve 200 g $NaC_2H_3O_2 \cdot 3H_2O$ in 800 ml water.

　　v. **Phenanthroline solution:** Dissolve 100 mg 1,10-phenanthroline monohydrate $(C_{12}H_8N_2 \cdot H_2O)$ in 100 ml water by stirring and heating to 80°C, without boiling. Discard the solution if it darkens or add 2 drops conc. HCl to avoid heating.

　　vi. **Stock iron solution:** Add 20 ml conc. H_2SO_4 to 50 ml water slowly and dissolve 1.404 g ferrous ammonium sulphate $[Fe(NH_4)_2(SO_4)_2 \cdot 6H_2O]$. Add 0.1 N $KMnO_4$ dropwise until a faint pink colour appears and persists. Dilute to 1 litre with distilled water. 1 ml = 200 µg Fe.

　　vii. Standard iron solutions: Take 5 and 50 ml stock iron solutions in volumetric flasks and dilute each to 1 litre. 1 ml = 1 and 10 µg Fe, respectively.

c. Procedure

　　i. **Total iron:** Take 50 ml of mixed sample in a 125 ml conical flask. If this volume is expected to contain more than 200 µg iron use a smaller portion and dilute to 50 ml. Add 2 ml conc. HCl, 1 ml NH_2OH, a few glass beads and heat to boiling till the volume is reduced to 15–20 ml. Cool the solution and transfer to a 50 ml volumetric flask. Add 10 ml $NH_4C_2H_3O_2$ buffer solution and 4 ml phenanthroline solution, dilute to 50 ml with distilled water. Mix thoroughly and allow

10–15 minutes for development of colour. Take spectrophotometer reading at 510 nm.

ii. **Dissolved iron:** Filter sample through a 0.45 µm membrane filter into a vacuum flask containing 1 ml conc. HCl/100 ml sample. Analyse as above and express as total dissolved iron.

iii. **Ferrous iron:** Acidify freshly collected sample with 2 ml conc. HCl/100 ml of sample. Take 50 ml water sample, add 20 ml phenanthroline solution and 10 ml $NH_4C_2H_3O_2$ solution and mix thoroughly. Measure the colour after 15 minutes.

iv. Calculate ferric iron by subtracting ferrous from total iron.

v. **Colour measurement:** Prepare a series of standards by accurately pipetting volumes of standard iron solution into 125 ml conical flask and diluting to 50 ml. Follow steps as in (i) and plot a calibration curve. Use weaker standard for measuring 1–10 µg iron.

d. Calculation

Calculate the iron content from the calibration curve.

$$\text{Fe in mg/L} = \frac{\text{Fe in µg in final volume}}{\text{Water sample in ml}}$$

6.12. ESTIMATION OF MAGNESIUM FORM TOTAL HARDNESS (TH) AND CALCIUM

Procedure

Determine the value for total hardness and Ca by disodium dihydrate salt of EDTA and calculate Mg by the formula:

$$\text{Mg in mg/L} = \frac{\text{TH} - 2.497 \times \text{Ca}}{4.115}$$

6.13. ESTIMATION OF MANGANESE (Mn) BY PERSULPHATE SPECTROPHOTOMETRIC METHOD

a. Apparatus

i. Spectrophotometer for use at 525 nm with a light path of 1 cm or greater.

b. Reagents

i. **Special reagent:** Dissolve 75 g $HgSO_4$ in 400 ml conc. HNO_3 and 200 ml distilled water. Add 200 ml 85% phosphoric acid and 35 mg silver nitrate. Dilute the cooled solution to 1 litre with distilled water.

ii. Solid ammonium persulphate [$(NH_4)_2S_2O_8$].

iii. **1 ml = 50 µg Mn standard manganese solution:** Dissolve 3.2 g $KMnO_4$ in distilled water and make up to 1 litre. Heat for several hours near the boiling point, cool and filter. Standardise against sodium oxalate by the procedure given below.

- Accurately weigh one 0.1 mg, several 100 to 200 mg samples of $Na_2C_2O_4$ and transfer them to 400 ml beakers.
- Add 100 ml distilled water to each beaker and stir to dissolve.
- Add 10 ml 1 + 1 H_2SO_4 to each beaker and heat rapidly to 90 to 95°C.
- Titrate quickly with $KMnO_4$ to slight pink end point. Do not allow temperature to fall below 85°C. If necessary, warm during titration.

- Run a blank on distilled water and H_2SO_4.
- Calculate normality:

$$\text{Normality of KMnO}_4 = \frac{\text{Na}_2\text{C}_2\text{O}_4 \text{ in g}}{(A - B) \times 0.06701}$$

where,

A = titrant for sample in ml

B = titrant for blank in ml

Average the results of several titrations and calculate volume of this solution necessary to prepare 1 litre of standard manganese solution as follows:

$$\text{KMnO}_4 \text{ (in ml)} = \frac{4.55}{\text{Normality of KMnO}_4}$$

- Add 2 to 3 ml conc. H_2SO_4 and $NaHSO_3$ solution dropwise to this volume of $KMnO_4$ until the permanganate colour disappears.
- Boil to remove excess SO_2, cool and dilute with distilled water to 1000 ml.
- Dilute this solution further with distilled water to measure small amounts of Mn.

iv. **Sodium oxalate:** Solid $Na_2C_2O_4$.

v. **Sodium bisulphite:** Dissolve 10 g $NaHSO_3$ in 100 ml distilled water.

vi. 30% Hydrogen peroxide (H_2O_2).

c. Procedure

i. Take a suitable volume of water sample containing 0.05 to 2.0 mg Mn in a 250 ml conical flask. Add 5 ml special reagent and one drop H_2O_2. Concentrate to 90 ml by boiling or dilute to 90 ml.

ii. Add 1 g $(NH_4)_2S_2O_8$ and boil for 1 minute and then cool under the tap. Dilute to 100 ml with distilled water.

iii. Prepare standards in the range of the sample concentration by treating various amounts of standard Mn-solution in the same manner as in (i), and (ii) above.

iv. Make spectrophotometric measurements of standards and sample at 525 nm against distilled water blank. Use light path of 1 cm for Mn in the range of 100–1500 µg/100 ml final volume. Plot standard calibration curve and determine Mn concentration in the final 100 ml volume from the standard curve.

d. Calculation

$$\text{Mn (in ml)} = \frac{\text{Mn in µg/100 ml in final volume}}{\text{Water sample in ml}}$$

6.14. ESTIMATION OF MERCURY BY MERCURY ANALYSER

The technique of mercury analysis is known as cold vapour atomic absorption spectrophotometry. It is based on the principle that mercury vapour (atoms) absorbs resonance radiations at 253.7 nm at room temperature.

The mercury analyser consists of

i. Low pressure mercury lamp emitting 253.7 nm resonance line

ii. An absorption cell

iii. A filter

iv. Vapour generation system

v. A detector with associated electronic devices

a. Reagents

i. **Potassium permanganate solution (1%):** Dissolve 1 g $KMnO_4$ in distilled water, add 10 ml conc. H_2SO_4 and dilute to 100 ml.

ii. **Sodium hydroxide solution (20%):** Dissolve 20 g NaOH in distilled water and dilute to 100 ml with distilled water.

iii. **Stannous chloride solution (20% in 10% HCl):** Dissolve 20 g AR grade $SnCl_2$ in 10 ml conc. HCl with warming, cool and dilute to 100 ml with distilled water.

iv. Sulphuric acid in 1:1 dilution.

v. **10% nitric acid:** Dilute 10 ml conc. HNO_3 to 100 ml with distilled water.

b. Standard mercury solution

i. Dissolve 0.1354 g $HgCl_2$ in 2% HNO_3 and dilute to 1000 ml with 2% HNO_3 (1 ml = 0.1 mg Hg = 100 µg Hg). Stock solution of Hg is to be maintained with 2% HNO_3 and 0.01% $K_2Cr_2O_7$ solution.

ii. Dilute 10 ml of 100 µg Hg solution to 100 ml with 2% HNO_3 solution. (1 ml = 10 µg Hg = 10,000 ng Hg)

iii. Dilute 10 ml of 10 µg Hg solution to 100 ml with 2% HNO_3 solution. (1 ml = 1 µg Hg = 1000 ng Hg).

iv. Dilute 10 ml of 1000 ng Hg solution to 100 ml with 2% HNO_3 solution. (1 ml = 0.1 µg Hg = 100 ng Hg)

c. Preparation of standard graph

i. Take aliquots of 1 ml, 2 ml, 3 ml, 4 ml, 5 ml and 6 ml from 100 ng Hg stock solutions in BOD bottles. These solutions contain 100, 200, 300, 400, 500 and 600 ng Hg, respectively.

ii. Add required volume of distilled water to maintain the total volume at 93 ml before adding any reagents in the BOD bottles.

iii. Add 5 ml of 10% HNO_3 acid solution in BOD bottle.

iv. Add 2 ml $SnCl_2$ solution and stopper the BOD bottles immediately. (Up to this stage the total volume in BOD bottle should be 100 ml.)

v. Keep this BOD bottle (mercury vapour generator) on magnetic stirrer and continue for 5 minutes.

vi. Keep a filter rod on open position and mode switch to 'Hold' position.

vii. Adjust absorbance 0 to 2 or transmission 0 to 100 by moving adjustment switches of the instrument as per manufacturer's instructions.

viii. After 5 minutes, start the pump and allow mercury vapours to purge through the BOD bottle (i.e. vapour generation bottle).

ix. Note absorbance/transmittances on a meter at 'Hold' position.

x. Plot a standard graph between concentration of Hg and absorbance/transmittances.

d. Determination of mercury in water

i. Take 50 ml of water sample in a BOD bottle.

ii. Add 43 ml distilled water, 5 ml of 10% HNO_3 acid solution and 2 ml $SnCl_2$ solution and stopper the BOD bottle immediately.

iii. Keep BOD bottle as prepared above on magnetic stirrer. Start the magnetic stirrer and continue for five minutes.

iv. Measure absorbance/transmittance after 5 minutes.

v. Carry out blank in a similar manner and deduct from sample reading.

vi. Find out the concentration of mercury from the standard graph.

Calculations

$$\text{Hg in mg/L} = \frac{\text{Hg in ng (in 50 ml) water sample}}{50}$$

6.15. ESTIMATION OF NITRATE

The nitrate present in water sample can be estimated by spectrophotometer in ultraviolet range (220 and 275 nm), visible range (410 nm) and by ion selective electrode.

6.15.1. Estimation of Nitrate by Spectrophotometric Method in Ultraviolet Range

a. Apparatus

An UV spectrophotometer for use at 220 nm and 275 nm with matched silica cells of 1 cm or longer light path.

b. Reagents

i. **Nitrate free water:** Double distilled or de-ionised water to prepare solutions.

ii. **Stock nitrate solution:** Dissolve 0.7218 g KNO_3 (dried in hot air oven at 105°C overnight and cooled in desiccator) in distilled water and dilute to 1 litre. Preserve with 2 ml of $CHCl_3$/L. 1 ml = 100 µg NO_3^-–N

iii. **Standard nitrate solution:** Dilute 100 ml of stock solution to 1000 ml with water, preserve with 2 ml of $CHCl_3$/L. 1 ml = 10 µg NO_3^-–N

iv. **1 N hydrochloric acid:** Add 83 ml conc. HCl to 850 ml of distilled water and dilute to 1 litre.

c. Procedure

i. Add 1 ml HCl to 50 ml clear/filtered water sample and mix thoroughly.

ii. Prepare calibration standards in the range of 0–7 mg NO_3^-–N/L by diluting to 50 ml. Add 1 ml of HCl to each solution and mix well.

Nitrate standard solution (ml)	1	2	4	7	10	15	20	25	30	35
NO_3^-–N/L in mg/L	0.2	0.4	0.8	1.4	2.0	3.0	4.0	5.0	6.0	7.0

iii. Set the spectrophotometer at zero absorbance or 100% transmittance for re-distilled water at wavelength of 220 nm. Use a wavelength of 220 nm to obtain NO_3^- reading and a wavelength of 275 nm to determine interference due to dissolved organic matter. Read absorbance for standards and water sample. Subtract 2 times the absorbance reading at 275 nm, from the reading at 220 nm to obtain absorbance due to NO_3^-. Prepare a standard curve by plotting absorbance due to NO_3^- against NO_3^-–N concentration of standards. Obtain sample concentrations directly from standard curve by using corrected sample absorbance. Multiply by 4.43 to get the concentration of NO_3^-.

If the reading at 275 nm is more than 10% of the reading at 220 nm, erroneous result is obtained. Estimate the NO_3^- by spectrophotometric method in visible range given below.

6.15.2. Estimation of Nitrate by Spectrophotometric Method in Visible Range

a. Reagents

    i. **Phenol disulphonic acid solution:** Dissolve 25 g of phenol in 150 ml conc. H_2SO_4. Heat for 2 hours, cool and store in a coloured bottle.

    ii. **KMnO$_4$ solution (0.05 N):** Dissolve 0.2 g of $KMnO_4$ in distilled water and dilute to 1000 ml.

    iii. **Standard Ag$_2$SO$_4$ solution:** Dissolve 4.4 g Ag_2SO_4 in distilled water and dilute to 1000 ml. (1 ml of Ag_2SO_4 solution = 1 mg Cl).

    iv. **Oxalic acid solution (0.05 N):** Dissolve 0.265 g of oxalic acid in 100 ml of distilled water.

    v. **Nitrate stock solution:** Dissolve 0.7218 g AR grade KNO_3 in distilled water and dilute to 1000 ml (1 ml of KNO_3 solution = 0.1 mg NO_3^-–N = 100 µg NO_3^-–N).

    vi. **Standard nitrate solution:** Dilute 10 ml of 0.1 mg NO_3^-–N to 100 ml with distilled water (1 ml = 10 µg NO_3^-–N).

    vii. **30% ammonia solution.**

b. Preparation of standard graph

    i. Take at least 3 aliquots of standard NO_3^-–N solution ranging from 0 to 10 µg in 100 ml volumetric flasks.

    ii. Determine the equivalent of chlorine, if present by precipitating chloride as AgCl by adding Ag_2SO_4.

    iii. Warm, cool and filter through Whatman No. 40 filter paper.

    iv. Add 2–5 drops of $KMnO_4$ solution to remove nitrites (NO_2) if present.

    v. Remove excess $KMnO_4$ colour by adding oxalic acid dropwise till colour disappears.

    vi. Add 2 ml phenol disulphonic acid solution.

    vii. Make the pH of the solution about 7 by adding ammonia solution. Stir thoroughly and make the volume 100 ml with distilled water.

    viii. Measure the absorbances by spectrophotometer at 410 nm.

    ix. Plot a graph between concentration and absorbance.

c. Procedure

    i. Take about 50 ml of water sample in a 250 ml beaker.

    ii. Add calculated volume of Ag_2SO_4 solution (as equivalent to chloride) to precipitate chlorides as AgCl. Warm the solution, cool and filter through Whatman No. 40 filter paper.

iii. Add 3–5 drops of $KMnO_4$ solution to remove nitrites (NO_2), if present.

iv. Remove excess $KMnO_4$ by adding oxalic acid dropwise.

v. Keep the beaker on hot water bath and allow the solution to evaporate completely.

vi. Remove the beaker and cool.

vii. Add 2 ml phenol disulphonic acid solution and 20–30 ml distilled water.

viii. Make the pH of the solution about 7 by adding ammonia solution. Stir thoroughly and make the volume 100 ml with distilled water.

ix. Measure the absorbance by spectrophotometer at 410 nm.

x. Determine the amount of $NO_3^- $–N in the water sample in µg/L.

Calculations

$$NO_3^- - N \text{ in } \mu g/L = \frac{NO_3^- - N \text{ in } \mu g}{\text{Volume of water sample in ml}} \times 1000$$

$$NO_3 = 4.43 \times NO_3^- - N$$

6.15.3. Estimation of Nitrate by Ion Selective Electrode

a. Apparatus

i. Ion meter

ii. Nitrate and reference electrodes

iii. Magnetic stirrer with TFE coated stirring bar

b. Reagents

i. **Nitrate free water:** Double distilled or de-ionised water to prepare solutions.

ii. **Stock nitrate solution:** Dissolve 0.7218 g, dried and cooled potassium nitrate (KNO_3) in water and dilute to 1 litre; 1 ml = 100 µg NO_3^-–N.

iii. **Standard nitrate solutions:** Dilute 1, 10, and 50 ml stock nitrate solutions to 100 ml each to obtain standards of 1, 10 and 50 mg NO_3^-–N/L, respectively.

iv. **Buffer solution:** Dissolve 17.32 g $Al_2(SO_4)_3 \cdot 18H_2O$, 3.43 g Ag_2SO_4, 1.28 g H_3BO_3, and 2.52 g sulfamic acid (H_2NSO_3H) in 800 ml water. Adjust to pH 3.0 by slowly adding 0.1 N NaOH. Dilute to 1000 ml and store in a dark glass bottle.

c. Procedure

i. Take 10 ml of 1 mg NO_3^-–N/L standard in a 50 ml beaker, add 10 ml buffer and stir with magnetic stirrer. Stop stirring and immerse electrodes. Start stirring again.

ii. Record millivolt reading when stable (after about 1 minute). Repeat with 10 and 50 mg NO_3^-–N/L standards.

iii. Plot potential measurement versus NO_3^-–N concentration on a semilogarithmic graph paper, potential measurement of the standards in millivolt on arithmetic scale and NO_3^-–N concentration on logarithmic scale. The calibration curve should be a straight line with a slope of +57 ± 3/decade at 25°C. Recalibrate the probes and the instruments several times every day using the 10 µg NO_3^-–N/L standard.

v. In case of direct reading ion meters, follow manufacturer's instructions to set up and calibrate the ion meter using standards in the prescribed range. In many cases standards diluted with the buffer are supplied with the ion meter by the manufacturer.

d. Calculation

Read nitrate nitrogen ($NO_3^- - N$) concentration of the sample from the calibration curve or directly from the meter. Multiply it by 4.43 to get the concentration of NO_3^-.

6.16. ESTIMATION OF OXYGEN (DISSOLVED)

Oxygen dissolved in water is important for aquatic organisms. It is normally estimated by titrimetric method.

a. Apparatus required

  i. DO sampler
 ii. Biochemical oxygen demand (BOD) bottles of 300 ml volume having narrow mouth, flared lip and with tapered and pointed ground glass stoppers.
iii. Siphon tube

b. Reagents

  i. **MnSO$_4$ solution:** Dissolve 480 g $MnSO_4 \cdot 4H_2O$ or 400 g $MnSO_4 \cdot 2H_2O$ or 364 g $MnSO_4 \cdot H_2O$ in distilled water, filter and make the volume 1 litre.
 ii. **Alkali-iodide-azide solution:** Dissolve 500 g NaOH and 135 g NaI (or 700 KOH and 150 g KI) in distilled water; add 10 g NaN_3 dissolved in 40 ml distilled water and make the volume 1 litre.
iii. **Conc. H$_2$SO$_4$**
 iv. **Starch indicator:** Dissolve 2 g laboratory grade soluble starch and 0.2 g salicyclic acid (preservative) in 100 ml hot distilled water and cool.
  v. **Standard sodium thiosulphate solution (0.025 M):** Dissolve 6.205 g $Na_2S_2O_3 \cdot 5H_2O$ in distilled water; add 1.5 ml 6 N NaOH or 0.4 g solid NaOH and dilute to 1000 ml.
 vi. **Standard potassium bi-iodate solution (0.0021 M):** Dissolve 812.4 mg $KH(IO_3)_2$ in distilled water and make the volume 1000 ml.

c. Determination of molarity (M) of sodium thiosulphate solution

  i. Take 100–150 ml distilled water in an Erienmeyer flask, add about 2 g KI and dissolve.
 ii. Add a few drops of conc. H_2SO_4 and 20 ml potassium bi-iodate solution.
iii. Dilute the mixture to 200 ml, add a few drops of starch indicator and titrate with sodium thiosulphate solution till first disappearance of blue colour.

Calculate the molarity of sodium thiosulphate solution as:

$$M = \frac{20 \times 0.025}{V}$$

where,
  V = Volume (in ml) of sodium thiosulphate solution used.

d. Procedure

  i. Drain out the liquid, if any, in the flared lip of the BOD bottle by the siphon tube.
 ii. Remove the stopper from the BOD bottle, add 1 ml $MnSO_4$ solution and 1 ml alkali-iodide-azide solution and mix well by inverting the stoppered bottle several times.

iii. Wait for a few minutes to settle the $Mn(OH)_2$ precipitate.

iv. Add 1 ml conc. H_2SO_4, restopper and mix thoroughly by inverting the stoppered bottle several times till the precipitate dissolves completely.

v. Add a few drops of starch indicator

vi. Titrate 201 ml of the solution against sodium thiosulphate solution till first disappearance of blue colour.

e. Calculation

$$\text{Dissolved oxygen (in mg/L)} = \frac{V \times M}{0.025}$$

where, V = ml of thiosulphate solution, and

        M = Molarity of thiosulphate solution.

6.17. ESTIMATION OF pH BY pH METER

a. Apparatus

i. pH meter with temperature compensation facility to read 0.1 pH unit within the range of 0 to 14.

ii. **Reference electrode:** Refer to manufacturer's manual regarding the use and care of the reference electrodes. In case of non-sealed electrodes, fill up with correct electrolyte to accurate level and make sure that the junction is properly-wet.

iii. **Glass electrode:** Follow manufacturer's instructions regarding use of electrode.

b. Reagents

i. **Potassium hydrogen phthalate buffer, 0.05 M, pH 4.00:** Dissolve 10.12 g $KHC_8H_4O_4$ in 1000 ml freshly boiled and cooled distilled water.

ii. **0.025 M potassium dihydrogen phosphate + 0.025 M disodium hydrogen phosphate buffer, pH 6.86:** Dissolve 3.387 g KH_2PO_4 + 3.533 g Na_2HPO_4 in 1000 ml freshly boiled and cooled distilled water.

iii. **0.01 M sodium borate decahydrate (borax buffer), pH = 9.18:** Dissolve 3.80 g $Na_2B_4O_7.10H_2O$ in 1000 ml freshly boiled and cooled distilled water.

iv. Store the buffer solutions in polyethylene bottles. Replace buffer solutions every 4 weeks.

Buffer tablets of pH 4.0, 7.0 and 9.2 are also available, which can be used as standards.

c. Procedure

i. Take out the electrodes from storage solution, rinse, blot dry with soft tissue paper, place in pH 4.00 buffer solution and standardise pH meter according to manufacturer's instructions.

ii. Remove electrodes from the first buffer, rinse thoroughly with distilled water, blot dry and immerse in pH 6.86 or 9.18 (or 7.0 or 9.2) buffer solutions. Read pH, which should be within 0.1 unit of the pH.

iii. Determine pH of the sample using the same procedure as in (ii) after establishing equilibrium between electrodes and sample.

iv. To ensure homogeneity, stir the sample gently while measuring pH.

6.18. ESTIMATION OF PHOSPHATE

Phosphate in water sample can be estimated by ascorbic acid and stannous chloride spectrophotometric methods.

6.18.1. Estimation of Phosphate by Ascorbic Acid Spectrophotometric Method

a. Apparatus

A spectrophotometer with infrared phototube for use at 880 nm or filter photometer provided with a red filter.

b. Reagents

i. **5 N Sulphuric acid:** Dilute 70 ml conc. H_2SO_4 to 500 ml with distilled water.

ii. **Potassium antimonyl tartrate solution:** Dissolve 1.3715 g $K(SbO)C_4H_4O_6 \cdot \frac{1}{2}H_2O$ in 400 ml distilled water and dilute to 500 ml.

iii. **Ammonium molybdate solution:** Dissolve 20 g $(NH_4)_6Mo_7O_{24} \cdot 4H_2O$ in 500 ml distilled water.

iv. **0.1 M Ascorbic acid:** Dissolve 1.76 g ascorbic acid in 100 ml distilled water. Keep it at 4 °C and discard after a week.

v. **Combined reagents:** Mix 50 ml 5 N H_2SO_4, 5 ml potassium antimonyl tartrate, 15 ml ammonium molybdate solution and 30 ml ascorbic acid solution in the given sequence at room temperature. Use it within 4 hours.

vi. **Stock phosphate solution:** Dissolve 219.5 mg anhydrous KH_2PO_4 in distilled water and dilute to 1000 ml. 1 ml = 50 µg PO_4^{3-}–P.

vii. **Standard phosphate solution:** Dilute 50 ml stock solution to 1 litre with distilled water. 1 ml = 2.5 µg P.

c. Procedure

i. **Sample treatment:** Take 50 ml water sample in a 125 ml conical flask, add 1 drop of phenolphthalein indicator. Remove any red colour by adding 5N H_2SO_4. Add 8 ml combined reagent and mix thoroughly.

ii. Wait for 10 minutes, measure absorbance of each sample at 880 nm within 30 minutes. Use reagent blank as reference.

iii. **Correction for turbid or coloured samples:** Prepare a sample blank by adding all reagents except ascorbic acid and potassium antimonyl tartrate to the sample. Subtract blank absorbance from sample absorbance reading.

iv. **Preparation of calibration curve:** Prepare calibration for a 1 cm light path from a series of standards between 0.15 to 1.30 mg P/L range. Use distilled water blank with the combined reagent.

v. Plot a graph with absorbance versus phosphate concentration which will be a straight line.

d. Calculation

$$PO_4 \text{ as mg P/L} = \frac{P \text{ (in mg from the calibration curve)} \times 100}{\text{Water sample in ml}}.$$

$$PO_4 = 3.0645 \times P$$

6.18.2. Determination of Phosphorus by Stannous Chloride: Spectrophotometric Method

a. Reagents

   i. **Ammonium molybdate solution:** Dissolve 25 g ammonium molybdate [$(NH_4)_6$ $(Mo_7)_{24}\cdot4H_2O)$] in 150 ml distilled water, add 280 ml conc. H_2SO_4, allow to cool and dilute to 1000 ml by adding distilled water.

   ii. **Stannous chloride solution:** Dissolve 2.5 g $SnCl_2$ in 100 ml glycerol by heating on water bath with constant stirring.

   iii. **Stock phosphate solution:** Dissolve 0.439 g KH_2PO_4 in distilled water and dilute to 1000 ml with distilled water (1 ml = 100 µg PO_4).

   iv. **Standard phosphate solution:** Dilute 10 ml, 100 µg PO_4 solution to 100 ml with distilled water (1 ml = 10 µg PO_4).

b. Preparation of standard graph

   i. Take suitable aliquots from standard phosphate solution in the range of 0–10 µg PO_4 in 100 ml volumetric flasks.

   ii. Add 4 ml ammonium molybdate solution in each flask and mix thoroughly.

   iii. Add 0.5 ml stannous chloride solution and mix thoroughly.

   iv. Dilute to 100 ml with distilled water and mix thoroughly.

   v. Measure absorbances/transmittances at 690 nm after 10 minutes.

   vi. Carry out blank in a similar manner.

   vii. Plot a graph between concentrations and absorbances/transmittances.

c. Procedure for samples

   i. Take suitable water sample in a 100 ml volumetric flask. (If turbidity is present, filter through 0.45 um filter.)

   ii. Add 4 ml ammonium molybdate solution, 0.5 ml stannous chloride solution and dilute to 100 ml with distilled water and mix thoroughly as was done in case of phosphate solution.

   iii. Measure absorbance/transmittance at 690 nm after 10 minutes.

   iv. Read the phosphorous concentration from standard graph.

d. Calculations

$$P \text{ in } µg/L = \frac{P \text{ in } µg \text{ in the sample}}{\text{Water sample in ml}} \times 1000$$

$$PO_4 = 3.0645 \times P$$

6.19. ESTIMATION OF POTASSIUM BY FLAME PHOTOMETER

a. Apparatus

   i. Direct reading type flame photometer.

   iii. Plastic bottles to store all solutions.

b. Reagents

   i. **Stock potassium solution:** Weigh 1.907 g KCl, dry at 140°C and cool in desiccator. Transfer it to 1 litre volumetric flask and dissolve in 1 litre distilled water. 1 ml of solution = 1 mg K.

ii. Intermediate potassium solution, dilute 10 ml of above stock potassium solution with water to 100 ml; 1 ml = 0.1 mg K.

iii. **Standard potassium solution:** Dilute 10 ml of above intermediate solution with water to 100 ml, 1 ml = 10 µg K.

c. Procedure

i. Follow the operating procedures described in section 3.8.5, *see* page 27 (estimation of K_2O by flame photometry)

ii. Prepare a blank and potassium calibration standards, in any of the applicable ranges, 0–100, 0–10, or 0–1 mg K/L. Set the instrument zero with standard containing no potassium. Measure the emission at 766.5 nm to find out the potassium content directly or prepare a calibration curve from which potassium concentration of the sample, or diluted sample, can be determined.

d. Calculation

K in mg/L = K in mg/L from the calibration curve × Dilution

where,

$$\text{Dilution} = \frac{\text{Water sample in ml} + \text{Distilled water in ml}}{\text{Water sample in ml}}$$

6.20. ESTIMATION OF SILICATE (SiO_2) BY AMMONIUM MOLYBDATE SPECTROPHOTOMETRIC METHOD

a. Apparatus

A spectrophotometer with 1 cm light path for use at 815 nm.

b. Reagents

i. 1 N sulphuric acid

ii. 1 + 1 hydrochloric acid

iii. **Ammonium molybdate reagent:** Dissolve 10 g $(NH_4)_6Mo_7O_{24}\cdot4H_2O$ in distilled water with stirring and gentle warming. Dilute to 100 ml with distilled water. Filter if necessary. Adjust pH between 7 and 8 with silica free NH_4OH or $NaOH$ solution and store in polyethylene bottle.

iv. **Oxalic acid solution:** Dissolve 7.5 g $H_2C_2O_4\cdot H_2O$ in distilled water and dilute to 100 ml.

v. **Stock silica solution:** Dissolve 313 mg sodium hexafluorosilicate (Na_2SiF_6) in 1000 ml distilled water. 1 ml = 0.1 mg SiO_2.

vi. **Standard silica solution:** Dilute 100 ml stock solution to 1000 ml; 1 ml = 10 µg SiO_2.

vii. **Reducing agent:** Dissolve 500 mg 1-amino-2-naphthol-4-sulphonic acid and 1 g Na_2SO_3 in 50 ml distilled water with gentle warming. Add a solution of 30 g $NaHSO_3$ in 150 ml distilled water. Filter into a plastic bottle. Discard when the solution becomes dark.

c. Procedure

i. To 50 ml water sample containing between 20 and 100 µg silica, add in rapid succession 1.0 ml 1 + 1 HCl and 2 ml ammonium molybdate reagent. Mix thoroughly and allow to stand for 5 to 10 minutes. Add 2 ml oxalic acid solution and mix thoroughly. Add 2 ml reducing agent within 2–15 minutes from the moment of adding oxalic acid and mix thoroughly.

ii. Read absorbance at 815 nm after 5 minute adjusting the instrument to zero absorbance using distilled water blank.

iii. Dilute 2, 4, 6, 8 and 10 ml silica working standard solution to 50 ml volumes and proceed as in (i), and (ii) above to prepare a calibration curve.

d. Calculation

Read silica content of sample from the calibration curve.

6.21. ESTIMATION OF SODIUM (Na) BY FLAME PHOTOMETER
a. Apparatus

i. Direct reading type flame photometer.

iii. Plastic bottles to store all solutions.

b. Reagents

i. **Stock sodium solution:** Take 2.542 g NaCl, dry at 140°C and cool in desiccator. Transfer it to 1 litre volumetric flask and dissolve in 1 litre distilled water. 1 ml solution = 1 mg Na.

ii. Intermediate sodium solution, dilute 10 ml of above stock sodium solution with water to 100 ml; 1 ml = 0.1 mg Na.

iii. **Standard sodium solution:** Dilute 10 ml of above intermediate solution with water to 100 ml, 1 ml = 10 µg Na.

c. Procedure

i. Follow the operating procedures described in section 3.8.5, *see* page 27 (estimation of Na_2O by flame photometry)

ii. Prepare a blank and sodium calibration standards in any of the applicable ranges, 0–100, 0–10, or 0–1 mg Na/L. Set the instrument zero with standard containing no sodium. Measure the emission at 589 nm to find out the sodium content directly or prepare a calibration curve from which sodium concentration of the sample, or diluted sample, can be determined.

d. Calculation

Na in mg/L = Na in mg/L from the calibration curve × Dilution
where,

$$\text{Dilution} = \frac{\text{Water sample in ml} + \text{Distilled water in ml}}{\text{Water sample in ml}}$$

6.22. ESTIMATION OF SULPHATE

The sulphate present in water sample can be estimated by spectrophotometric, gravimetric and nephelometric methods.

6.22.1. Estimation of Sulphate by Spectrophotometric Method

a. Reagents

i. **Conditioning reagent:** Add 75 g of NaCl, 30 ml of conc. HCl, and 100 ml of ethyl alcohol to 300 ml of distilled water. Add 50 ml of glycerol to the solution.

ii. **Standard solution:** Dissolve 0.1479 g of anhydrous Na_2SO_4 in 1 litre of distilled water. This solution contains 100 mg/L of sulphate. Prepare solutions of different strengths by dilution with distilled water.

iii. $BaCl_2$ crystal

b. Procedure

i. Add 1 ml of conditioning reagent and a pinch of $BaCl_2$ crystal to 20 ml of the standard solution and stir well.

ii. Take the standard solutions in spectrophotometer at 420 nm and standardize the spectrophotometer with at least 3 standards of different strengths.

iii. Determine the actual amount of the sulphate from the spectrophotometer reading.

6.22.2. Estimation of Sulphate by Gravimetric Method

a. Reagents

i. 1:1 dil. HCl

ii. $BaCl_2$ solution (10%)—dissolve 10 g of $BaCl_2$ in 100 ml of distilled water.

iii. Methyl red indicator (0.1%)—dissolve 0.1 g of methyl red in 100 ml distilled water.

iv. $AgNO_3$ solution (1%)—dissolve 1 g of $AgNO_3$ in 100 ml distilled water.

b. Procedure

i. Filter about 100 ml of water sample through sintered glass crucible of G-4 porosity by vacuum pump.

ii. Add 2 ml of dil. HCl and evaporate to dryness.

iii. Ignite at 180°C to burn out organic matter, if present.

iv. Take out the mass with 2 ml dil. HCl.

v. Boil and cool so that insoluble materials will settle to the bottom.

vi. Dilute with 100 ml distilled water, boil, cool and filter through Whatman No. 40 filter paper.

vii. Take the filtrate in a 250 ml beaker.

viii. Adjust pH to 4–5 by adding dil. HCl, which can be ascertained by methyl red indicator.

ix. Add 2 ml of dil. HCl in excess and boil on a burner.

x. Warm $BaCl_2$ solution and add 5 ml of it to the boiling solution.

xi. Continue boiling for 15 minutes; $BaSO_4$ will precipitate out.

xii. Allow the precipitate to settle overnight.

xiii. Filter $BaSO_4$ through Whatman No. 42 filter paper.

xiv. Wash the residue with hot water till complete exclusion of chloride ions. This can be checked by adding $AgNO_3$ solution (white precipitation will take place in the presence of the chloride ions).

xv. Ignite the $BaSO_4$ residue at 900–950°C in a muffle furnace.

xvi. Cool the residue and weigh $BaSO_4$.

c. Calculation

$$\text{Sulphate in mg/L} = \frac{\text{Weight of } BaSO_4 \text{ in mg} \times 411.6}{\text{Volume of water sample in ml}}$$

6.22.3. Estimation of Sulphate (SO_4) by Nephelometry

a. Apparatus

i. Nephelometric turbidity meter with sample cells for use at 420 nm with a light path of 2.5 to 10 cm.

ii. Magnetic stirrer.

iii. Stopwatch with measuring facility in seconds.

b. Reagents

i. **Buffer solution-A:** Dissolve 30 g magnesium chloride ($MgCl_2 \cdot 6H_2O$), 5 g sodium acetate ($CH_3COONa \cdot 3H_2O$), 1 g potassium nitrate (KNO_3), and 20 ml acetic acid CH_3COOH (99%) in 500 ml distilled water and make up to 1000 ml.

ii. **Buffer solution-B:** This solution is necessary when SO_4^{2-} concentrations in water sample is less than 10 mg/L. Prepare as buffer solution A and add 0.111 g sodium sulphate (Na_2SO_4).

iii. Barium chloride ($BaCl_2$) crystals of 20 to 30 mesh size.

iv. **Standard sulphate solution:** Dilute 10.4 ml standard 0.02 N H_2SO_4 into 100 ml (1 ml = 100 µg SO_4^{2-}).

v. **Approximately 0.05 N standard sodium carbonate solution:** Dry 3 to 5 g sodium carbonate (Na_2CO_3) at 250°C for 4 hours and cool in a desiccator. Accurately weigh 2.5 ± 0.2 g to the nearest mg, dissolve in distilled water and make to 1 litre.

vi. **Approximately 0.1 N standard H_2SO_4:** Dilute 2.8 ml conc. sulphuric acid to 1 litre. Standardise against 40 ml 0.05 N Na_2CO_3 with about 60 ml distilled water in a beaker by titrating potentiometrically to pH 5. Lift out electrodes, rinse into the same beaker and boil gently for 3 to 5 minutes under a watch glass cover. Cool to room temperature, rinse cover glass into the beaker and titrate to pH 4.3. Calculate normality of sulphuric acid by the following formula.

$$\text{Normality (N)} = \frac{A \times B}{53 \times C}$$

where,

A = wt. of Na_2CO_3 (in g) used to make 1 litre standard solution at (v) above

B = Volume of Na_2CO_3 (in ml) taken for standardisation titration at (vi) above

C = Volume of acid (in ml) used in standardisation titration

In case, potentiometric titration is not possible, use bromcresol green indicator to complete the titration. The indicator is prepared by dissolving 100 mg bromcresol green sodium salt in 100 ml distilled water.

vii. **Standard sulphuric acid, 0.02N:** Dilute the approximate 0.1 N solution to 1 L. Calculate volume to be diluted as:

$$\text{Volume in ml} = \frac{20}{N}$$

where,

N = Exact normality of the approximate 0.1 N solution.

c. Procedure

i. Standardise nephelometer following manufacturer's instructions.

ii. Measure the turbidity of sample-blank, a water sample in which $BaCl_2$ is not added.

iii. Take 100 ml water sample (or a suitable portion made up to 100 ml) in a 250 ml conical flask. Add 20 ml buffer solution A and mix. While stirring, add a spoonful of $BaCl_2$ crystals. Stir for 60 ± 2 seconds.

iv. Measure turbidity of the sample at 5 ± 0.5 minutes after stirring stopped.

v. Prepare SO_4^{2-} standards at 5 mg/L increments in the range of 0 to 40 mg/L SO_4^{2-} according to the following procedure.

SO_4^{2-} (in mg/L)	5	10	20	30	40
Standard SO_4^{2-} solution (in ml)	5	10	20	30	40
Distilled water (in ml)	95	90	80	70	60

vi. Develop $BaSO_4$ turbidity for the standards as mentioned in (iii), and (iv) above.

vii. Draw calibration curve between turbidity and SO_4^{2-} concentration (in mg/L)

viii. In case buffer solution B is used for samples containing less than 10 mg/L SO_4^{2-}, run a reagent-blank with distilled water in place of sample, developing turbidity and reading it as in (iii) and (iv) above.

d. Calculation

i. If buffer solution A is used, read SO_4^{2-} concentration for the water sample from the calibration curve after subtracting the turbidity of sample-blank from the turbidity of the treated sample. If less than 100 ml sample is used, multiply the result by 100/sample volume (in ml).

ii. If buffer solution B is used read SO_4^{2-} concentration in the treated sample from the calibration curve after subtracting the turbidity of sample-blank from the turbidity of the treated sample. Then read SO_4^{2-} concentration for the reagents from the turbidity value of the reagent-blank (see procedure 'viii') from the calibration curve. Record the corrected SO_4^{2-} concentration in the sample after subtracting the reagent-blank SO_4^{2-} concentration from the sample SO_4^{2-} concentration.

6.23. ESTIMATION OF SULPHIDE BY IODOMETRIC METHOD

Excess of standard iodine solution is added in the acidic media to get oxidized sulphide. Residual iodine is titrated against standard sodium thiosulphate solution using starch as an indicator and sulphide is calculated accordingly.

a. Reagents

   i. **Standard iodine solution (0.01 N):** Dissolve 10–15 g KI in small volume of distilled water and add 1.28 g iodine crystals. Allow to dissolve completely. Dilute to 1000 ml with distilled water. This solution should be standardised against standard sodium thiosulphate solution using starch as an indicator.

   ii. **Standard sodium thiosulphate solution (0.01 N):** Dissolve 2.482 g sodium thiosulphate in distilled water and dilute to 1000 ml with distilled water.

   iii. 6 N HCl.

   iv. **Starch solution:** Prepare about 5% starch solution with boiling water.

b. Standardization of iodine solution

Take suitable aliquot of iodine solution in a 250 ml conical flask. If necessary, dilute with distilled water. Add 2 ml 6 N HCl solution and mix thoroughly. Titrate ~~~~~ ~~~~~~ ~~~ sodium thiosulphate solution. Add 2–3 ml starch solution near the end point and continue titration till blue colour disappears. Calculate normality (N) by following formula.

$$N_{(Iodine)} = \frac{N\ (thio.) \times Volume\ of\ thio.}{Volume\ of\ iodine\ solution}$$

c. Procedure for samples

   i. Take suitable aliquot of iodine in a 250 ml conical flask. Dilute with distilled water if necessary.

   ii. Add 2 ml of 6N HCl solution.

   iii. Pipette out a suitable amount of water sample and discharge it under a surface of solution present in conical flask. Observe the colour of iodine which must be present at this stage, if not add more standard iodine solution so that the colour of iodine remains brown.

   iv. Titrate the content of conical flask against standard sodium thiosulphate solution using starch as an indicator. Disappearance of blue colour is the end point.

d. Calculation

$$S^{2-}\ in\ mg/L = \frac{\left(Volume\ of\ iodine\ in\ ml \times N_{iodine}\right) - \left(volume\ of\ thio. \times N_{thio.}\right)\ \times 16{,}000}{Volume\ of\ water\ sample}$$

(1 ml 1N iodine solution = 16 mg S^{2-})

6.24. ESTIMATION OF TOTAL ACIDITY

If the water sample is suspected to be acidic, i.e. pH is less than 7, total acidity is to be determined.

a. Apparatus

Standard laboratory glassware such as burettes, volumetric flasks and beakers.

b. Reagents

   i. 0.02 N NaOH solution prepared by dissolving 0.8 g of NaOH in 1000 ml distilled water.

   ii. Phenolphthalein indicator solution

c. Procedure

Take 10 ml of water sample in a beaker and add 3 drops of phenolphthalein indicator. Titrate with 0.02 N NaOH till light pink end point is reached.

d. Calculation

Total acidity in mg/L = volume of the titrant (in ml) × 100

6.25. ESTIMATION OF TOTAL ALKALINITY

Alkalinity of the water sample of pH ≥ 4.5 can be determined by titration with HCl and H_2SO_4.

6.25.1. Titration with HCl

a. Apparatus

Standard laboratory glassware such as burettes, volumetric flasks and beakers.

b. Reagents

i. 0.02 N HCl prepared by diluting 1.72 ml conc. HCl to 1000 ml.
ii. Methyl orange indicator

c. Procedure

Take 10 ml of water sample in a beaker and add 3 drops of methyl orange indicator. Titrate with 0.02 N HCl till light pink end point is reached.

d. Calculation

Total alkalinity (as $CaCO_3$ in mg/L) = Volume of the titrant (in ml) × 100.

6.25.2. Titration with H_2SO_4

a. Apparatus

Standard laboratory glassware such as burettes, volumetric flasks and beakers.

b. Reagents

i. **Standard sodium carbonate, approximately 0.05 N:** Dry 3 to 5 g Na_2CO_3, at 250°C for 4 hours and cool in a desiccator. Accurately weigh 2.5 ± 0.2 g, dissolve in distilled water and make to 1000 ml.
ii. **Standard H_2SO_4, 0.1 N:** Dilute 2.8 ml conc. H_2SO_4 to 1 litre. Standardise against 40 ml 0.05 N Na_2CO_3 with about 60 ml distilled water, in a beaker by titrating potentiometrically to pH 5. Lift out electrodes, rinse into the same beaker and boil gently for 3 to 5 minutes under a watch glass cover. Cool to room temperature, rinse cover glass into beaker and finish titration to pH 4.3. Calculate normality of sulphuric acid by the following formula.

$$\text{Normality, N} = \frac{A \times B}{53 \times C}$$

where,
  A = Weight of Na_2CO_3 used in g
  B = Volume of Na_2CO_3 solution taken for standardisation titration in ml
  C = Volume of acid used in standardisation titration in ml

In case, potentiometric titration is not possible use bromcresol green indicat r t complete the titration.

iii. **Standard sulphuric acid, 0.02 N:** Dilute the approximate 0.1 N solution to 1 litre. Calculate volume to be diluted as:

$$\text{Volume in ml} = \frac{20}{N}$$

where, N = Exact normality of the 0.1 N solution.

iv. **Bromcresol green indicator, pH 4.5:** Dissolve 100 mg bromcresol green sodium salt in 100 ml distilled water.

c. Procedure

Take about 10 ml of water sample in a beaker and add 2 to 3 drops of bromcresol green indicator. Titrate until change in colour to blue (pH 4.9) or yellow (pH 4.3) is observed. Note the volume of titrant (in ml) used.

d. Calculation

$$\text{Total alkalinity (as } CaCO_3 \text{ in mg/L)} = \frac{D \times N \times 50000}{\text{Volume of water sample in ml}}$$

where,

D = Volume (in ml) of titrant

N = Normality of titrant

6.25.3. If the Alkalinity of the Water Sample Seems to be More, Phenolphthalein (pH = 8.3) is to be Used

a. Reagent

Phenolphthalein indicator solution: Dissolve 5 g phenolphthalein in 500 ml 95% ethyl alcohol and add 500 ml distilled water.

b. Procedure

Take 25 to 50 ml water sample in a conical flask. Add 2 to 3 drops of phenolphthalein indicator. If it turns pink (pH >8.3), titrate with 0.02 N H_2SO_4 till the colour disappears.

c. Calculation

$$\text{Phenolphthalein alkalinity (as } CaCO_3 \text{ in mg/L)} = \frac{E \times N \times 50000}{\text{Volume of water sample in ml}}$$

where,

E = Volume (in ml) of titrant used to phenolphthalein pink end point

N = Normality of titrant

6.26. ESTIMATION OF TOTAL SOLIDS (TS), SUSPENDED SOLIDS (SS) AND TOTAL DISSOLVED SOLIDS (TDS) IN WATER SAMPLE

The water sample invariably contains some solids, which is known as total solids (TS). The total solid is divisible into suspended solids (SS), which can be separated by filtration and total dissolved solids (TDS), which accounts for the soluble chemicals present in water. The TDS is more significant than other two types of solids as far as water quality is concerned.

6.26.1. Total Solids

i. Take 50 ml of water sample in a pre-weighted (W_1) platinum dish.

ii. Evaporate the water on a hot plate at 105°C.

iii. Heat, cool and weigh the dish several times till the weight becomes constant (W_2).

Calculation

Total solids (in mg/L) = $(W_2 - W_1) \times 20$

6.26.2. Suspended Solids

i. Filter 50 ml of water sample through a pre-weighted (W_1) sintered glass crucible of G-4 porosity by vacuum pump.

ii. Dry the sintered glass crucible in an oven at 105°C.

iii. Cool the sintered glass crucible and weigh.

iv. Heat, cool and weigh the crucible several times till the weight becomes constant (W_2).

v. Amount of suspended solid = $W_2 - W_1$

Calculation

Suspended solids (in mg/L) = $(W_2 - W_1) \times 20$

6.26.3. Total Dissolved Solids

i. Filter 50 ml of water sample through Whatman No.40 filter paper into a pre-weighted (W_1) platinum dish.

ii. Evaporate the water on a hot plate at 105°C.

iii. Heat, cool and weigh the dish several times till the weight becomes constant (W_2).

Calculation

Dissolved solids (in mg/L) = $(W_2 - W_1) \times 20$

During heating of the water sample, the volatile solids escape as a result of which, amount of total dissolved solid is underestimated. Since the total dissolved solid is proportional to the electrical conductivity of water, the amount of total dissolved solid can be estimated by the following equation.

Total dissolved solids (TDS in mg/L) = 0.64 × Electrical conductivity (EC in µS/cm)

6.27. ESTIMATION OF TOTAL HARDNESS

Total hardness is determined by EDTA titrimetry and expressed in terms of $CaCO_3$ in mg/L. Two similar methods are described below, one without complexing agent and another with the use of complexing agent.

6.27.1. Estimation of Total Hardness (TH) by EDTA Titrimetry without Complexing Agent

a. Reagents

i. **Ammonia buffer solution:** Take 142 ml of NH_3 solution and add 17.5 g of NH_4Cl and make up to 250 ml with distilled water.

ii. **Erichrome Black T solution:** Dissolve 0.4 g of Erichrome Black T in 100 ml of distilled water and add 95% ethanol and make the volume 1000 ml.

iii. **EDTA solution:** Add 3.723 g of disodium dihydrate salt of EDTA and 2 NaOH pellets to 1000 ml of distilled water. Standardise against standard Ca solution (1 ml = 1 mg $CaCO_3$).

b. Procedure

i. Take 10 ml of water sample in a beaker, add 3 drops of ammonia buffer solution and 2 drops of Erichrome black T solution.

ii. Titrate against EDTA till sky blue end point is reached.

c. Calculation

Total hardness as $CaCO_3$ in mg/L = Volume of EDTA in ml × 100

6.27.2. Estimation of Total Hardness (TH) by EDTA Titrimetry with Complexing Agent

a. Reagents

i. **Buffer solution:** Dissolve 16.9 g NH_4Cl in 143 ml conc. NH_4OH. Add 1.25 g magnesium salt of ethylenediaminetetraacetate (EDTA) and dilute to 250 ml with distilled water. Store it in a plastic bottle. The solution is to be used within 1 month from the date of preparation.

   If the Mg salt of EDTA is unavailable, dissolve 1.179 g disodium salt of ethylenediaminetetraaceticacid dihydrate (AR grade) and 780 mg magnesium sulphate ($MgSO_4\cdot7H_2O$) or 644 mg magnesium chloride ($MgCl_2\cdot6H_2O$) in 50 ml distilled water. Add this solution to 16.9 g NH_4Cl and 143 ml conc. NH_4OH and dilute to 250 ml with distilled water. To attain the highest accuracy, adjust to exact equivalence through appropriate addition of a small amount of EDTA or $MgSO_4$ or $MgCl_2$.

ii. **Complexing agent:** Add 250 mg magnesium salt of 1,2-cyclohexanediaminetetra acetic acid per 100 ml water sample only if presence of interfering ions is suspected and sharp end point is not obtained.

iii. **Indicator Eriochrome Black T sodium salt:** Dissolve 0.5 g salt in 100 ml triethanolamine or 2-ethylene glycol monomethyl ether. It can also be used in dry powder form by grinding 0.5 g salt with 100 g NaCl.

iv. **Standard EDTA titrant, 0.01M:** Dissolve 3.723 g disodium dihydrate salt of EDTA in distilled water and dilute to 1000 ml. Store in a polyethylene bottle.

v. **Standard calcium solution:** Take 1 g anhydrous $CaCO_3$ in a 500 ml flask and add 1+1 HCl slowly through a funnel till all $CaCO_3$ is dissolved. Add 200 ml distilled water and boil for a few minutes to expel CO_2, cool and add a few drops of methyl red indicator and adjust to the intermediate orange colour by adding 3 N NH_4OH or 1 + 1 HCl, as required. Dilute to 1000 ml with distilled water, 1 ml = 1 mg $CaCO_3$.

b. Procedure

i. Dilute 25 ml sample to 50 ml with distilled water. Add 2 ml buffer to give a pH of 10.0 to 10.1; 5 ml of complexing agent and 2 drops of indicator solution. Titrate with EDTA titrant till the colour changes from reddish tinge to blue. For best results, select a sample volume that needs less than 15 ml EDTA titrant and complete the titration within 5 minutes after addition of buffer solution.

c. Calculation

$$\text{Total hardness CaCO}_3 \text{ in mg/L} = \frac{A \times B \times 1000}{\text{Water sample in ml}}$$

where, A = EDTA in ml, B = wt (in mg) of $CaCO_3$ equivalent to 1 ml EDTA titrant

6.28. DETERMINATION OF TURBIDITY BY TURBIDITY METER/NEPHELOMETER

Turbidity in water occurs due to presence of suspended matter. It is measured by turbidity meter or nephelometer. Turbidity measurement is based on comparison of the intensity of light scattered by a sample with that of standard reference suspension under similar conditions.

a. Reagents

i. **Stock hydrazine sulphate solution:** Dissolve 1 g hydrazine sulphate $[(NH_2)_2H_2SO_4]$ in water and dilute to 100 ml with distilled water.

ii. **Stock hexamethylene tetramine solution:** Dissolve 10 g hexamethylene tetramine $[(CH_2)_6 N_4]$ in distilled water and dilute to 100 ml with distilled water.

iii. **Standard mixture suspension solution:** Mix 5 ml of stock hydrazine sulphate solution and 5 ml of hexamethylene tetramine solution in 100 ml volumetric flask and dilute to 100 ml with distilled water (Turbidity = 400 NTU). For less than 400 NTU, dilute 10 ml 400 NTU solution to 100 ml in a volumetric flask. (Turbidity = 40 NTU).

iv. **Preparation of standards:** Take suitable aliquots of standard suspension turbidity solution ranging from 0 – 40 NTU in 100 ml volumetric flasks and make up to the mark.

b. Calibrate the instrument as per manufacturer's instructions

The calibration method is available in instruction manual of the manufacturer of the instrument. Turbidity can be determined by both turbidity meter and nephelometer. It varies with instrument and manufacturing company.

c. Measurement of turbidity in samples

Place the sample in absorbing tube of the meter and compare with the standards. (*Note*: Dilution may be done in case of reading is out of calibration range).

d. Calculations

$$\text{Turbidity (in NTU)} = \frac{A \times (B + C)}{C}$$

where,

A = Turbidity of diluted sample

B = Volume of dilution water in ml

C = Water sample analysed in ml

Turbidity readings may be reported as follows:

Reading range NTU	Report to the nearest NTU
0–1.0	0.05
1–10	0.10
10–40	1
40–100	5
100–400	10
100–400	10
400–1000	50
>1000	100

6.29. DATA VALIDATION

Check if

i. $\dfrac{\text{Total cations} - \text{total anions}}{\text{Total cations} + \text{total anions}} < 0.1$ Concentrations in mEq/L

ii. $\dfrac{\text{Na}^+ \text{ in mEq/L}}{\text{Cl}^- \text{ in mEq/L}} = 0.8 - 1.2$

iii. $\dfrac{\text{TDS in mg/L}}{\text{EC in } \mu \text{ mho/cm}} = 0.55 - 0.9$

iv. If pH <8.3 phenolphthalein alkalinity, $CaCO_3$ in mg /L = 0 ?

If the above criteria are satisfied, then the analyses are correct, otherwise the analyses are to be repeated.

6.30. CHEMICAL EQUIVALENCE

Concentration in milliequivalet per liter (mEq/L) = (Concentration in mg/L) × conversion factor

$$\text{Conversion factor} = \frac{\text{Valency}}{\text{Formula weight}}$$

For Na^+, conversion factor is $\dfrac{1}{23} = 0.0435$, For Ca^{2+}, conversion factor is $\dfrac{2}{40} = 0.05$

Conversion factors of common cations and anions are given in Table 6.2.

Table 6.2: Conversion factors of common cations and anions

Cation	Conversion factor	Anion	Conversion factor
Calcium (Ca^{2+})	0.05000	Bicarbonate (HCO_3^-)	0.01639
Iron (ferric) (Fe^{3+})	0.05372	Carbonate (CO_3^{2-})	0.03333
Iron (ferrous) (Fe^{2+})	0.03581	Chloride (Cl^-)	0.02821
Magnesium (Mg^{2+})	0.08226	Fluoride (F^-)	0.05264
Manganese (Mn^{2+})	0.03640	Nitrate (NO_3^-)	0.01613
Potassium (K^+)	0.02557	Phosphate (PO_4^{3-})	0.03159
Sodium (Na^+)	0.04350	Sulphate (SO_4^{2-})	0.02082

6.31. GRAPHICAL REPRESENTATION OF WATER CHEMISTRY

Water chemistry data are presented with the help of different diagrams. Some common diagrams are pi-diagram, Collin's bar diagram, Stiff pattern diagram and Schoeller diagram. In pi-diagram, the concentrations of cations and anions in mEq/L are converted to 180° and plotted in half circle, cation above and anion below. One such diagram with Ca^{2+}, 44.51 mg/L; Mg^{2+},12.04 mg/L; Na^+, 21.02 mg/L; K^+, 5.63 mg/L; Fe, 0.70 mg/L; Cl^-,32.4 mg/L; HCO_3^-,156.41 mg/L; SO_4^{2-},11.48 mg/L; NO_3, 16.55 mg/L and F^-, 0.28 mg/L is shown in Fig. 6.5. In modified Collin's bar diagram (Fig. 6.6), the percentages of cations and anions are plotted side by side. In Stiff pattern diagram (Fig. 6.7), the mEq/L values of cations (to the left) and anions (to the right) are represented by horizontal lines from a common zero line at the center. Other ends of the lines are joined sequentially to form a definite pattern. The Schoeller diagram (Fig. 6.8) represents the concentrations of cations and anions. Gibbs diagrams show relationship between water chemistry and aquifer lithology. Plots $(Na^+ + K^+)/$ $(Na^+ + K^+ + Ca^{2+})$ versus total dissolved solids (TDS) (Fig. 6.9) and $Cl^-/(Cl^- + HCO_3^-)$ versus total dissolved solids (TDS) (Fig. 6.10) show whether the groundwater chemistry is dominantly influenced by 'evaporation', 'rock composition' or 'precipitation'. Piper trilinear diagram (Fig. 6.11) is used to show the chemical relationship among the cations and anions and to determine the water type. The lower left and right triangles are used to plot the cation and anion percentages, which are projected onto the diamond-shaped field. Plot in field-5 indicates that carbonate hardness exceeds 50%, i.e. chemical properties of water is dominated by alkaline earths and weak acids whereas plot in field-6 suggests that noncarbonate hardness exceeds 50%. Plot in field-7 indicates that chemical properties of the water are dominated by alkalies and strong acids

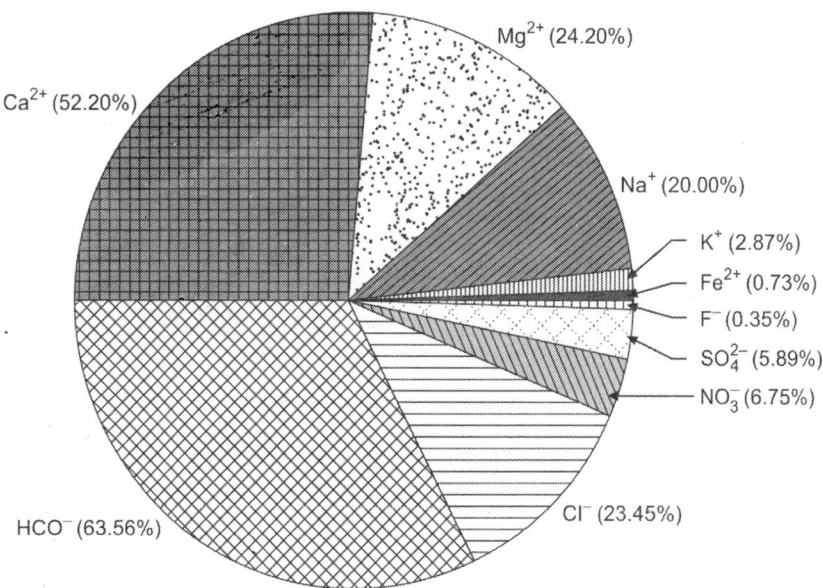

Fig. 6.5: Pi diagram

while plot in field-8 suggests dominance of primary alkalinity and carbonates. Mixed waters, i.e. when none of the cation and anion pairs exceeds 50% plot in field-9.

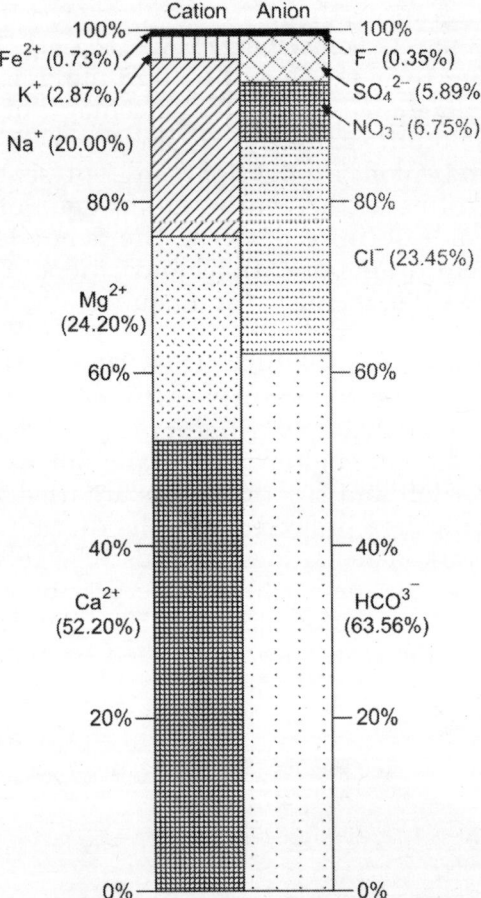

Fig. 6.6: Modified Collin's bar diagram

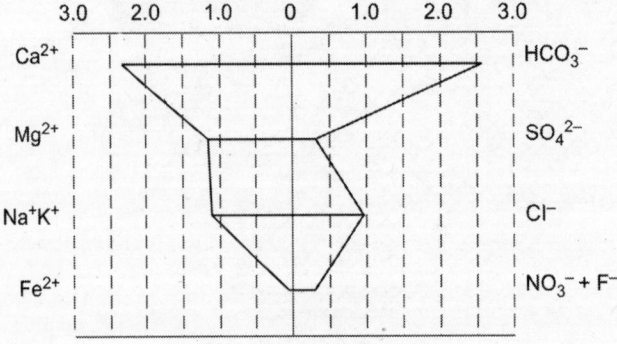

Fig. 6.7: Stiff pattern diagram

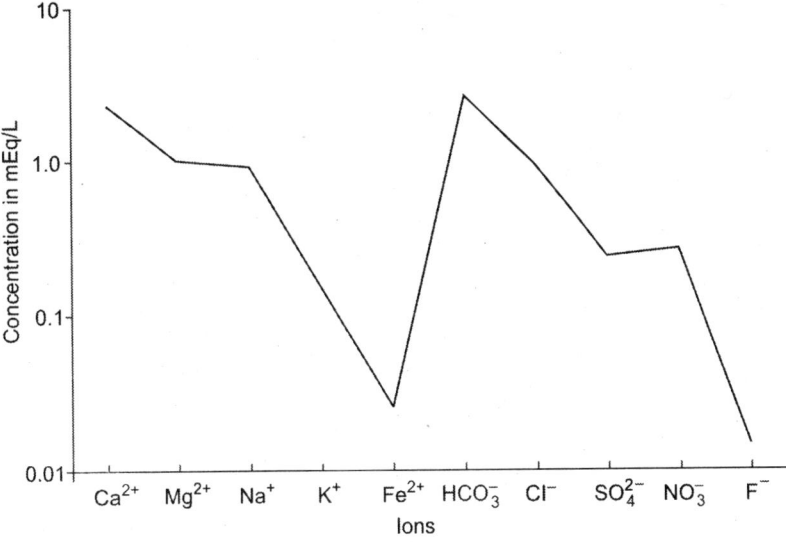

Fig. 6.8: Schoeller diagram

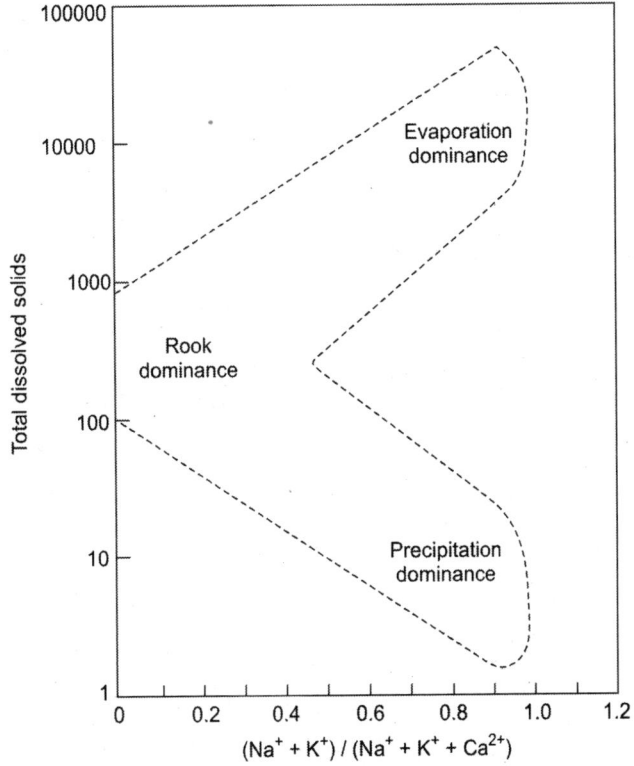

Fig. 6.9: Gibbs diagram for cations

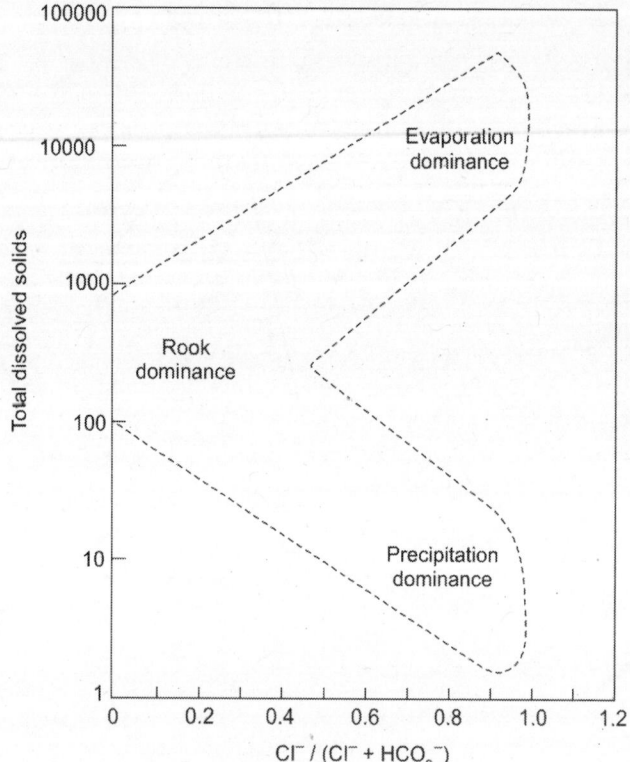

Fig. 6.10: Gibbs diagram for anions

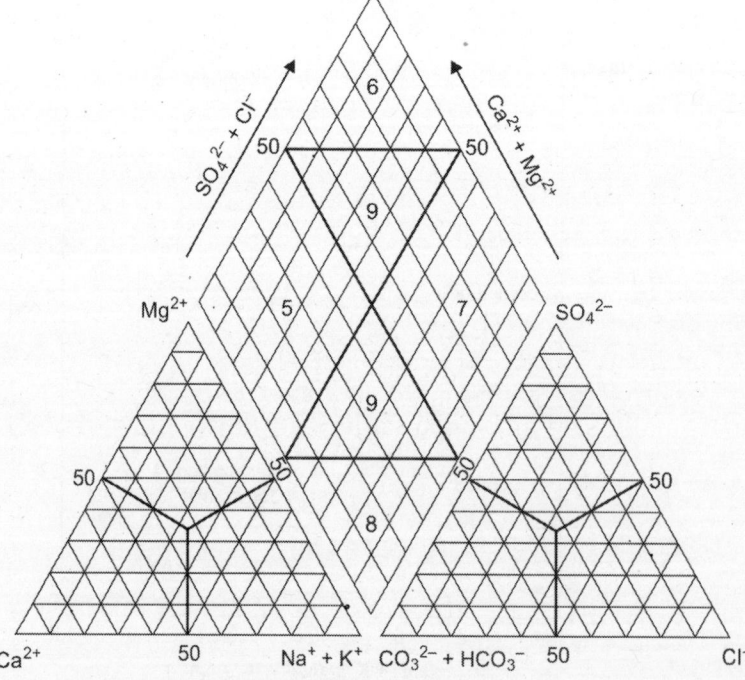

Fig. 6.11: Piper trilinear diagram

6.32. SUITABILITY OF WATER IN DIFFERENT FIELDS OF USE

Water is commonly used for drinking, irrigation and in industries. The specifications of waters in these fields are different from each other.

6.32.1. Drinking Water

The acceptable and permissible limits of different constituents recommended by the Bureau of Indian Standards (BIS, 2012) are given in Table 6.3. There are two limits, viz. acceptable and permissible. Water with parameters below the acceptable limits is good for human consumption. Since, in many instances, these limits are exceeded due to different reasons, upper limits, known as permissible limits have been fixed by Bureau of Indian Standards and World Health Organisation.

Hardness of water is due to the presence of divalent metallic cations of which Ca^{2+} and Mg^{2+} are most important. These ions react with soap to form precipitates making the water unsatisfactory for household cleansing. The hardness classification by Sawyer and Mc Carty (1967) is given in Table 6.4.

6.32.2. Irrigation Water

Suitability of water for irrigation depends upon several factors out of which quality of water, soil type and character, salt tolerance of plants, climate and drainage are important. Presence of soluble salts in excess amounts can cause harm to the crops. Total dissolved solids of values 0–480 mg/L, 480–1760 mg/L and more than 1760 mg/L can cause no problem, increasing problem and severe problem for plant growth respectively.

Chemical parameters like sodium adsorption ratio (SAR), residual sodium carbonate (RSC), permeability index (PI), percent sodium (% Na), potential soil salinity (PS), Kelley's ratio (KR) and magnesium ratio (MR) determine the suitability of water for irrigation use.

6.32.2.1. Sodium adsorption ratio (SAR)

The sodium adsorption ratio (SAR) is calculated by the formula

$$SAR = \frac{Na^+}{\sqrt{(Ca^{2+} + Mg^{2+})/2}}$$

where concentrations are in mEq/l.

Water is classified as excellent, good, medium and bad depending upon the SAR values 0–10, 10–18, 18–26 and more than 26, respectively.

6.32.2.2. Residual sodium carbonate (RSC)

Residual sodium carbonate (RSC) is calculated by the formula $RSC = (CO_3^{2-} + HCO_3^-) - (Ca^{2+} + Mg^{2+})$, where all concentrations are expressed in mEq/l. The water is classified into safe, marginally safe and unsuitable with RSC values <1.25, 1.25–2.5 and >2.5, respectively.

6.32.2.3. Permeability index (PI)

The permeability index (PI) is calculated by the formula:

$$PI = \frac{Na^+ + \sqrt{HCO_3^-}}{Na^+ + Ca^{2+} + Mg^{2+}} \times 100$$

where all concentrations are expressed in mEq/L.

Table 6.3: Extract from drinking water standard formulated by the Bureau of Indian Standards (BIS) IS 10500–2012

Characteristics	Acceptable limit	Permissible limit
Physical and chemical standards		
1. Turbidity (units on NTU scale)	1.0	5.0
2. Colour (units on Hazen scale)	5.0	15
3. Taste and odour	Agreeable	Agreeable
4. pH	6.5–8.5	No relaxation
5. Total alkalinity as $CaCO_3$, mg/L	200	600
6. Total hardness as $CaCO_3$, mg/L	200	600
7. Total dissolved solids, mg/L	500	2000
8. Aluminium as Al, mg/L	0.03	0.2
9. Ammonia, mg/L	0.5	No relaxation
10. Anionic detergents, as MBAS, mg/L	0.2	1.0
11. Arsenic as As, mg/L	0.01	No relaxation
12. Barium as Ba, mg/L	0.7	No relaxation
13. Boron as B, mg/L	0.5	1.0
14. Cadmium as Cd, mg/L	0.003	No relaxation
15. Calcium as Ca, mg/L	75	200
16. Chloride as Cl, mg/L	250	1000
17. Chromium, as hexavalent Cr, mg/L	0.05	No relaxation
18. Copper, as Cu, mg/L	0.05	1.5
19. Cyanide as CN, mg/L	0.05	No relaxation
20. Fluorides as F, mg/L	1.0	1.5
21. Iron as Fe, mg/L	1.0	No relaxation
22. Lead as Pb, mg/L	0.01	No relaxation
23. Magnesium as Mg, mg/L	30	100
24. Manganese as Mn, mg/L	0.1	0.3
25. Mercury as Hg, mg/L	0.001	No relaxation
26. Mineral oil, mg/L	0.5	No relaxation
27. Molybdenum as Mo, mg/L	0.07	No relaxation
28. Nickel as Ni, mg/L	0.02	No relaxation
29. Nitrates as NO_3, mg/L	45	No relaxation
30. Phenolic compounds as phenol, mg/L	0.001	0.002
31. Selenium as Se, mg/L	0.01	No relaxation
32. Silver as Ag, mg/L	0.1	No relaxation
33. Sulphate as SO_4, mg/L	200	400
34. Zinc as Zn, mg/L	5.0	15
Radioactivity		
35. Gross alpha emitters, Bq/L	0.1	No relaxation
36. Gross beta emitters, Bq/L	1.0	No relaxation

Table 6.4: Hardness classification of water by Sawyer and McCarty (1967)

Hardness (mg/L as CaCO₃)	Water class	Hardness (mg/L as CaCO₃)	Water class
0–75	Soft	150–300	Hard
75–150	Moderately hard	>300	Very hard

Irrigation waters are classified into class 1, class 2 and class 3 by plotting of permeability index (PI) and total dissolved solids (in mEq/L) in Doneen's diagram (Fig. 6.12).

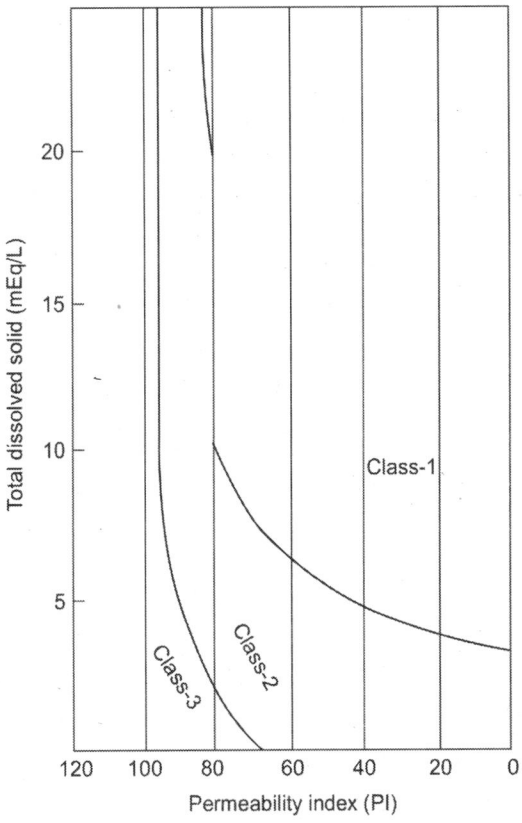

Fig. 6.12: Doneen's diagram

6.32.2.4. Percent sodium (%Na)

Sodium concentration is important in classifying irrigation water as sodium reacts with soil to reduce its permeability. Sodium concentration is usually expressed in terms of percent sodium (%Na) and is calculated by the formula:

$$\% Na = \frac{(Na^+ + K^+) \times 100}{Na^+ + K^+ + Ca^{2+} + Mg^{2+}}$$

where all concentrations are expressed in mEq/l.

Irrigation waters are classified into five types on the basis of sodium percent. These are excellent, good, permissible, doubtful and unsuitable with %Na up to 20, 20–40,

40–60, 60–80 and more than 80. Wilcox proposed a diagram (Fig. 6.13) in which sodium percentage is plotted against electrical conductivity for determining the suitability of water for irrigation.

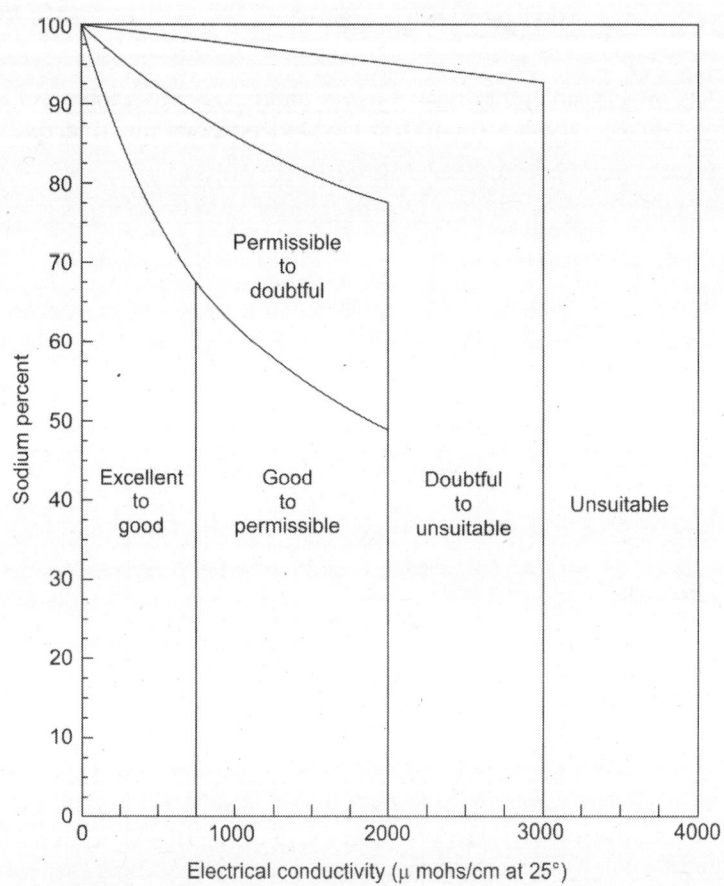

Fig. 6.13: Wilcox diagram

6.32.2.5. Potential soil salinity (PS)

The potential soil salinity (PS) is calculated by the sum of chloride and half of sulphate ions, i.e. $PS = Cl^- + \frac{1}{2} SO_4^{2-}$, where the concentrations are expressed in meq/l. Irrigation waters are divided into three classes, viz. excellent to good, good to injurious and injurious to unsuitable on the basis of potential soil salinity values of less than 5, 5–10 and more than 10, respectively.

6.32.2.6. Kelley's ratio

Kelley et al (1940) suggested that the sodium problem in irrigation water can be conveniently expressed on the basis of Kelley's ratio which is computed by the formula:

$$\text{Kelley's ratio} = \frac{Na^+}{Ca^{2+} + Mg^{2+}}$$

where all concentrations are expressed in meq/l.

Water having Kelley's ratio more than one is considered unsuitablet for irrigation.

6.32.2.7. Magnesium ratio (MR)

The magnesium ratio is calculated by the formula:

$$MR = \frac{Mg^{2+} \times 100}{Mg^{2+} + Ca^{2+}}$$

where all concentrations are expressed in mEq/L.
Magnesium ratio causes harmful effect to soil when it exceeds 50.

6.32.2.8. US salinity diagram

The United States Salinity Laboratory Diagram (Fig. 6.14) prepared from the values of electrical conductivity and sodium absorption ratio (SAR) classify irrigation waters into good, moderate and bad waters.

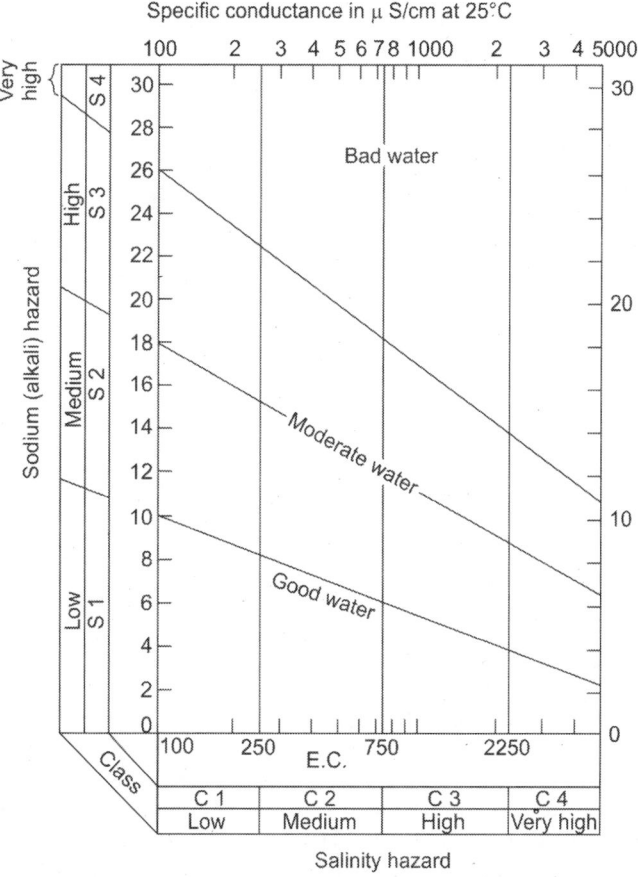

Fig. 6.14: United States Salinity Laboratory Diagram

6.32.3. Quality of Water used in Industries

The quality requirements of waters used in different industries vary widely; for example, water of very low quality like sea water can be satisfactorily used for cooling of condensers, whereas, waters for high-pressure boilers must meet stringent criteria. The specifications of water used for washing, cooling and manufacturing processes are widely different. Water quality may affect the product by biological activity, staining, corrosion, chemical reaction and contamination; deteriorate the equipment by corrosion, erosion, scale formation and may reduce the efficiency of the equipments. Highly pure water is required for the manufacture of pharmaceuticals and in paper industry. Different water parameters have different industrial implications, some of which are presented in Table 6.5.

Table 6.5: Effects of different water parameters in industrial use

Parameters	Probable effects
pH	pH 7 is required for most industries; low pH increases corrosion of concrete
Total dissolved solids	Causes foaming in boilers; affects colour and taste of finished products. High TDS leads to corrosion
Iron	Recommended value for food processing unit is 0.2 mg/L, for paper and photographic industry is 0.1 mg/L, less than 0.1 mg/L for cooling
Chloride	Causes corrosion of steel and aluminium
Fluoride	Harmful in industries producing pharmaceuticals and medical items
Calcium	Causes spots on films, forms scale, precipitates curds; interfere in formation of emulsions and upsets fermentation process and electroplating
Sulphate	Increases corrosiveness of water, low sulphate (less than 20 mg/L) is recommended for sugar industries
Nitrate	Injurious to dyeing of wool and silk fabrics, harmful in fermentation process for food and beverages

The corrosive property can be quantified with respect to corrosive ratio (CR), which is calculated by the formula:

$$\text{Corrosive ratio} = \frac{\dfrac{Cl^-}{35.5} + \dfrac{SO_4^{2-}}{48}}{\dfrac{(CO_3^{2-} + HCO_3^-)}{50}}$$

where, all ionic concentrations are expressed in mg/L.

Water samples with corrosive ratio more than 1.00 are generally unsuitable for industrial use. Incrustation is formed due to precipitation of the chemical constituents present in the water. It is of two type, soft incrustation and hard incrustation. Soft incrustation is formed due to precipitation of HCO_3^- of Ca and Mg and can be easily removed by acids and other chemicals. Hard incrustation is formed due to sulphates and silicates of Ca and Mg which are insoluble in acids and other chemicals and thus, cannot be removed.

Analysis of Soil Sample

Soil is the uppermost layer of the earth crust, which has been formed by different weathering processes. Most of the plants grow on soil from which they take water and nutrients. The type and quality of soil is dependent upon its physical and chemical characteristics. The physical parameters are particle size and the water holding capacity of the soil. The chemical parameters are pH, cation exchange capacity as well as presence and quantity of soluble salts, organic carbon, calcium carbonate, phosphorous, sulphur, nitrogen, etc. Soil testing involves sampling of soil and analysis of different physical and chemical parameters.

7.1. SAMPLING OF SOIL

The main objective of sampling is to collect representative samples of the entire area. If there is no considerable differences in colour, texture, elevation, etc. the whole area can be considered as a single unit, otherwise, the area is divided into a number of sampling units. In the later case, a number of samples are collected covering the entire area. Usually soil samples are taken from depths ranging from 15 to 22 cm from the surface of top soil. Collected samples are air dried, properly labeled and carefully stored avoiding contamination. It is often necessary to divide the original sample into two separate parts so that one can be used for physical tests and other can be crushed into desired size, either by hammering or by mechanical means followed by coning and quartering for analysis of chemical parameters.

7.2. DETERMINATION OF PHYSICAL PARAMETERS
7.2.1. Estimation of Surface Texture of Soil

Textural analysis indicates the percentage of solid material contained in different size fractions. It can be ascertained by mechanical and gravimetric methods.

7.2.1.1. Mechanical Method

a. Procedure

    i. Dry a soil sample of 30–70 gram at 105°C in an oven.

    ii. Stack sieves with 2 mm pore size at the top followed by 1 mm, 0.50 mm, 0.25 mm, 0.125 mm and 0.063 mm and a pan to catch materials finer than 0.063 mm at the bottom.

    iii. Place the stacked sieves with weighed sample in the topmost sieve with the lid on the topmost sieve and pan below the bottom-most sieve in sieve shaker.

    iv. Run the sieve shaker for 10–15 minutes.

    v. Stop the sieve shaker and collect the amounts of materials retained in each sieve.

    vi. Weigh the materials retained in each sieve and determine the percentage of each fraction by the following formula:

$$\text{Percentage of each fraction of soil} = \frac{\text{Weight of each fraction}}{\text{Total weight of the sample taken}} \times 100$$

Descriptive name of each fraction			
Size of particle	Grain name	Size of particle	Grain name
>2 mm	Granule	0.50–0.125 mm	Medium sand
2–1 mm	Very coarse sand	0.125–0.063 mm	Fine sand
1–0.50 mm	Coarse sand	< 0.063 mm	Silt and clay

7.2.1.2. Gravimetric Method

a. Reagents

    i. **Hydrogen peroxide:** 20 ml of 30% H_2O_2 diluted to 100 ml with distilled water.

    ii. **2N Hydrochloric acid:** 83 ml conc. HCl diluted to 500 ml with distilled water.

    iii. **2N Sodium hydroxide solution:** Dissolve 40 g NaOH in water and dilute to 500 ml with distilled water.

    iv. **1% Silver nitrate solution:** Dissolve 1 g $AgNO_3$ in water and dilute to 100 ml with distilled water.

b. Procedure

    i. Take 10 g of less than 2 mm size soil sample in a 500 ml beaker.

    ii. Add 125 ml distilled water and boil for about 10 minutes. Allow to cool and settle the suspension.

    iii. Decant supernatant and discard it.

    iv. Add 20 ml H_2O_2 solution and digest on water bath adding more H_2O_2 solution till no froth is evolved.

    v. Add about 20–30 ml 2N HCl and dilute with 100 ml distilled water.

    vi. Allow to stand for about one hour with intermittent stirring to make the soil free from carbonates.

    vii. Filter through Whatman No. 41 filter paper. Wash the residue with distilled water till it is free from chloride ion, which can be tested with $AgNO_3$ solution. Transfer the residue of soil into beaker.

    viii. Add 5 ml 2N NaOH solution and shake for about 30 minutes.

    ix. Transfer the content into a 1000 ml measuring cylinder and make up the volume to 1000 ml with distilled water.

    x. Stir the above content thoroughly for 1 minute and allow to stand.

    xi. After 4 minutes, take 25 ml of the mixture at 10 cm depth by pipette in pre-weighted platinum dish or in pre-weighted beaker.

    xii. Evaporate the mixture at 105°C in oven, cool and weigh. The difference in weight (W_1) is contributed by silt + clay.

xiii. Stir the remaining content in measuring cylinder thoroughly for 1 minute and allow to stand.

xiv. After 6 hours, take 25 ml of the mixture at 10 cm depth by pipette in pre-weighted platinum dish or in pre-weighted beaker.

xv. Evaporate the mixture at 105°C in oven, cool and weigh after it is dried completely. The difference in weight (W_2) is the weight of clay alone.

c. Calculation

Let wt. of sample = W_0 gram; Wt. of clay alone = W_2 gram; Wt. of silt + clay = W_1 gram

$$\% \text{ of clay} = \frac{W_2 \times 1000 \times 100}{W_0 \times 25} = \frac{W_2 \times 4000}{W_0}$$

$$\% \text{ of silt} = \frac{(W_1 - W_2) \times 1000 \times 100}{W_0 \times 25} = \frac{(W_1 - W_2) \times 4000}{W_0}$$

% of sand = 100 – (percent of silt + percent of clay).

7.2.2. Estimation of pH of Soil

Procedure

i. Take 10 gram of soil sample in a 100 ml beaker. Add 50 ml distilled water.

ii. Mix and intermittently stir for about 30 minutes.

iii. Measure the pH of the suspension using a pH meter.

Note: The pH meter can be standardized using standard pH solutions 0f 4.0, 7.0 and 9.2.

7.2.3. Estimation of Electrical Conductivity of Soil

Electrical conductivity is a measure of current carrying capacity of the soil. It gives an idea about the soluble salts present in it.

a. Procedure

i. Take 10 gram soil sample in a 100 ml beaker and add 50 ml distilled water to it.

ii. Stir for about 1 hour with intervals.

iii. Measure the conductivity of above suspension using standardised conductivity meter.

iv. Conductivity meter can be standardised using KCl (AR) of suitable concentration.

b. Calculation

Conductivity ($\mu s/cm$ or mhos/cm) = Cell constant × Meter reading (or as per the instructions of instrument manufacturer).

7.2.4. Estimation of Water Holding Capacity of Soil

Water holding capacity of soil depends upon its physical and chemical nature. When the soil is absolutely saturated with water so that water fills all the pores between particles without any air in the interspaces, the soil is said to be at its maximum water holding capacity.

a. Procedure

i. Dry the soil sample in oven at 105°C.

ii. Place a Whatman No.1 filter paper at the bottom of a perforated rectangular box.

iii. Take weight of the empty box (W_0).

iv. Fill the perforated box with the dry soil sample.

v. Take weight of the box with filled in soil (W_1).

vi. Place the box in a petri dish/tray.

vii. Fill water in the petri dish/tray just below the edges of the perforated box.

viii. Allow to stand as such for 12 hours to saturate the soil by water entering through holes.

ix. Take the box out after 12 hours.

x. Dry the outside surface of the box using tissue paper.

xi. Take weight of the box with saturated soil (W_2).

b. Calculation

$$\text{Water holding capacity of the soil (in \%)} = \frac{(W_2 - W_1) - (W_1 - W_0)}{(W_1 - W_0)} \times 100$$

7. 3. DETERMINATION OF CHEMICAL PARAMETERS

7.3.1. Estimation of Organic Carbon Content of Soil

For determination of organic carbon (matter) in soil samples, the sample is digested with excess of potassium dichromate solution in presence of sulphuric acid and excess of dichromate is determined by titrating against ferrous ammonium sulphate solution.

a. Reagents

i. **Potassium dichromate solution (1N):** Dissolve 49 g $K_2Cr_2O_7$ in distilled water and dilute to 1000 ml with distilled water.

ii. **Sulphuric acid with silver sulphate (5%):** Dissolve 5 g silver sulphate (Ag_2SO_4) in 100 ml conc. H_2SO_4.

iii. **Phosphoric acid:** 85%.

iv. **Ferrous ammonium sulphate solution (1N):** Dissolve 392.2 g ferrous ammonium slulphate {Fe(NH$_4$)$_2$(SO$_4$)·6H$_2$O} in 20 ml conc. H_2SO_4 and dilute to 1000 ml with distilled water.

v. **Barium diphenylamine sulphonate indicator 0.5%:** Dissolve 0.5 g barium diphenylamine sulphonate in water and dilute to 100 ml with distilled water.

b. Procedure

i. Take about 1.0 g soil sample in a 500 ml conical flask. Moisten the soil with few ml of distilled water.

ii. Add 10 ml potassium dichromate (1N) solution using pipette and shake well to mix it.

iii. Add 20 ml sulphuric acid—silver sulphate mixture, dilute to about 200 ml with distilled water and allow to stand for 30 minutes.

iv. Add 10 ml phosphoric acid, mix thoroughly and cool.

v. Add 1 ml barium diphenylamine sulphonate indicator.

vi. Titrate the above solution with 1 N ferrous ammonium sulphate solution till brilliant green colour end point is reached.

vii. Carry out a blank titration in similar manner without soil.

c. Calculation

$$\text{Organic carbon (in \%)} = \frac{(V_1 - V_2) \times N \times 0.003 \times 100}{W}$$

$$\text{Total organic matter} = \text{Percent of carbon} \times 1.724$$

Where,

W— Weight of soil sample

V_1—Blank titre value

V_2—Titre value of the sample

N—Normality of $K_2Cr_2O_7$ (here it is 1)

Note:

1. 1 ml 1N $K_2Cr_2O_7$ corresponds to 0.003 g organic carbon.
2. The factor 1.724 is based on the assumption that carbon constitutes 58% of organic matter.

7.3.2. Estimation of Cation Exchange Capacity of Soil

For the determination of cation exchange capacity, 1 N Sodium acetate solution of pH 8.2 is used. In this technique, sodium replaces all the exchangeable cations of soil when leached thoroughly and excess sodium ions are removed using ethyl alcohol. These sodium ions can be easily determined using flame photometer.

a. Reagents

i. **Sodium acetate solution** (IN, pH = 8.2): Dissolve 40.0 g sodium hydroxide in distilled water. Add 58 ml acetic acid and dilute to 1000 ml with distilled water. Adjust pH to 8.2 using NaOH.

ii. **Ethyl alcohol:** 95%

iii. **Ammonium acetate** (IN, pH = 7.0): Mix 58 ml acetic acid and 68 ml conc. ammonium hydroxide solution and dilute to 1000 ml. Adjust pH to 7.0 using NH_4OH or acetic acid.

b. Procedure

i. Take 5 g of soil powder in a 50 ml centrifuge tube. Add 30 mI 1N sodium acetate solution and shake for 5 minutes.

ii. Centrifuge the suspension till supernatant is clear. Decant the supernatant and discard.

iii. Repeat the above process thrice using 30 ml sodium acetate solution every time.

iv. Sodium saturated sample is then washed thrice with 30 ml of 95% ethyl alcohol with shaking for 5 minutes each time.

v. Wash with 20 ml ethyl alcohol till the supernatant shows conductivity between 40–55 μs/cm.

vi. The absorbed sodium is replaced from the sample with 30 ml volume of 1 N ammonium acetate solution. Repeat thrice.

vii. Decant the supernatant in a 100 ml volumetric flask. Make up to the mark with ammonium acetate solution.

viii. Determine sodium using flame photometer.

ix. Carry out a blank in a similar manner.

c. Calculation

$$\text{Cation exchange capacity (mEq/100 g)} = \frac{A \times V \times 100 \times 100}{W \times 23 \times 1000} = \frac{A \times V \times 10}{23 \times W}$$

Where,

A = Na content in ammonium acetate extract

V = Volume of extract in ml

W = Wt. of soil sample taken in gram

7.3.3. Estimation of Exchangeable Calcium and Magnesium in Soil

Calcium and magnesium can be determined by complexometric titration with EDTA using Patton and Reeder's reagent and Eriochrome Black-T as indicators at pH 12 and 10, respectively.

a. Reagents

i. **Concentrated hydrochloric acid**

ii. **Concentrated nitric acid**

iii. **EDTA solution 0.01 M:** Dissolve 3.722 g disodium dehydrate salt of EDTA in distilled water and dilute to 1000 ml with distilled water.

iv. **Standard zinc acetate solution (0.01 M):** Dissolve 0.2195 g zinc acetate (AR) in water and dilute to 100 ml with distilled water.

v. **Buffer solution (pH = 10):** Dissolve 67.5 g NH_4Cl in 510 ml ammonia solution and dilute to 1000 ml with distilled water.

vi. **Buffer solution (pH = 12):** Dissolve 200 g KOH in water and dilute to 1000 ml with distilled water.

vii. **Buffer solution (pH = 5.5):** Dissolve 200 g ammonium acetate in 30 ml acetic acid and dilute to 1000 ml with distilled water.

viii. **Patton and Reeder's reagent:** Dissolve 0.1 g in 100 ml methanol or water.

ix. **Eriochrome black-T:** Dissolve 0.1 g in 100 ml methanol or water.

x. **Xylenol orange solution:** Dissolve 0.1 g xylenol orange in 100 ml methanol or distilled water.

xi. **Hydroxylamine hydrochloride solution:** 5% in distilled water.

xii. **Triethanolamine solution:** 5% in distilled water.

b. Procedure

i. Take suitable aliquot from ammonium acetate extract from cation exchange capacity solution in a 250 ml beaker (Method 7.3.2).

ii. Evaporate to dryness on hot plate.

iii. Add 5 ml conc. HCI and dilute to 20 ml with distilled water and 1ml conc. HNO_3. Allow to complete the reaction.

iv. Evaporate to dryness on hot plate.

v. Extract with 1 ml conc. HCl, wash with distilled water and warm to dissolve. Filter through Whatman No. 40 filter paper if any residue is present.

vi. Make up the volume of filtrate to 100 ml in a volumetric flask.

vii. Add 2 ml 5% hydroxylamine hydrochloride solution and triethanolamine solution to suppress interfering cations like Fe, Mn, Al, etc.

c. Determination of calcium and magnesium

i. Take suitable aliquot of above solution (7.3.3. vii) in a 250 ml beaker.

ii. Dilute to about 100 ml with distilled water, add 20–25 ml buffer solution (pH = 10) and 2–5 drops of Eriochrome Black-T indicator.

iii. Titrate against 0.01 M EDTA solution.

iv. Peacock blue colour shows the end point of the reaction.

v. Run blank in the same manner and deduct from sample titration.

d. Determination of calcium only

i. Take suitable aliquot from the above solution (7.3.3.vii) in a 250 ml beaker.

ii. Dilute to about 100 ml with distilled water, add 20–25 ml buffer solution (pH = 12) and 2–5 drops of Patton and Reeder's reagent indicator solution.

iii. Titrate against 0.01 M EDTA solution.

iv. Peacock blue colour shows the end point of the reaction.

v. Carry out blank in a similar manner and deduct from sample titration.

e. Standardization of EDTA against zinc acetate solution

i. Take suitable aliquot of standard zinc acetate solution (0.01 M) in a 250 ml beaker.

ii. Dilute to 100 ml with distilled water, add 20–25 ml buffer solution (pH = 5.5) and 2–5 drops of xylenol orange indication solution.

iii. Titrate against 0.01 M EDTA solution.

iv. Golden yellow colour shows the end point of the reaction.

f. Calculations

Calculate molarity of EDTA from standard formula $M_1 \times V_1 = M_2 \times V_2$

Volume of EDTA for (Ca + Mg) – volume of EDTA for Ca = Volume of EDTA for Mg.

Exchangeable calcium: (mEq/100 g soil) =

$$\frac{\text{Volume of EDTA} \times \text{Molarity of EDTA} \times \text{Volume of aliquot taken for titration} \times 0.014 \times 100}{\text{Volume of original aliquot} \times \text{Weight of solil sample} \times 20 \times 20.04}$$

Exchangeable magnesium (mEq/100 g soil) =

$$\frac{\text{Volume of EDTA for Mg} \times \text{Molarity of EDTA} \times \text{Volume of aliquot taken for titration} \times 100}{\text{Volume of original aliquot} \times \text{Weight of solil sample} \times 12 \times 15}$$

(1 ml 1M EDTA = 40.8 mg of Ca = 24.3 mg of Mg).

7.3.4. Estimation of Exchangeable Sodium and Potassium in Soil

a. Reagents

Ammonium acetate (1N) pH = 7: Dissolve 68 ml ammonium hydroxide solution in 58 ml acetic acid and dilute to 1000 ml with distilled water. Adjust pH to 7.

b. Procedure

i. Take 5 g soil sample in 100 ml conical flask. Add 25 ml 1N ammonium acetate buffer solution. Shake for about 5 minutes.

ii. Filter through Whatman No. 41 filter paper.

iii. Use this filtrate for the determination of amounts of sodium and potassium by flame photometer.

Calculations

Exchangeable sodium (mEq/100 g soil) =

$$\frac{\text{Concentration from flame photometer} \times \text{Volume of aliquot taken} \times 100}{\text{Weight of solil sample} \times 23}$$

Exchangeable potassium (mEq/100 g soil) =

$$\frac{\text{Concentration from flame photometer} \times \text{Volume of aliquot taken} \times 100}{\text{Weight of solil sample} \times 19}$$

7.3.5. Estimation of Calcium Carbonate Content of Soil

a. Reagents

i. **Sulphuric acid solution (0.4 N):** Dilute 11.12 ml of concentrated sulphuric acid and make up to 1000 ml with distilled water.

ii. **Sodium hydroxide solution (0.4 N):** Dissolve 16 g of sodium hydroxide in water and dilute to 1000 ml with distilled water (Solution is to be standardised against standard acid solution).

iii. **Phenolphthalein indicator:** Dissolve 0.1 g in 100 ml distilled water.

iv. **Methyl orange:** 0.1 % solution in water.

b. Standardization of sulphuric acid solution

i. Take 0.1–0.2 g sodium carbonate in a 250 ml beaker, add 100 ml distilled water to dissolve it. Add 2–3 drops of methyl orange.

ii. Titrate against sulphuric acid solution.

iii. Purple colour shows the end point of the reaction.

$$\text{Normality of } H_2SO_4 = \frac{\text{Wt. of } Na_2CO_3 \times 1000}{53 \times \text{Volume of } H_2SO_4}$$

c. Procedure

i. Take about 1–2 g soil sample in a 250 ml flask. Add 25 ml 0.4 N H_2SO_4 solution.

ii. Dilute to 100 ml with distilled water.

iii. Boil for 5–10 minutes.

iv. Cool and titrate against standard NaOH solution using phenolphthalein as an indicator.

v. Carry out blank in a similar manner.

d. Calculation

% $CaCO_3$ =

$$\frac{[\text{Volume of NaOH for blank} - \text{Volume of NaOH for soil}] \times N_{NaOH} \times 0.05 \times 100}{\text{Weight of soil}}$$

(1 ml 1N H_2SO_4 = 0.05 g of $CaCO_3$ = 1 ml 1N NaOH).

7.3.6. Estimation of Calcium Bicarbonate Content of Soil

a. Procedure

i. Take the same solution from which calcium carbonate was estimated.

ii. Add 2–4 drops of methyl orange indicator.

iii. Titrate against same sulphuric acid solution till solution becomes red.

iv. Carry out blank in a similar manner and deduct from volume required for determination of $Ca(HCO_3)_2$.

Calculation

% $Ca(HCO_3)_2$ =

$$\frac{[\text{Volume of NaOH for } Ca(HCO_3)_2 - \text{Volume of NaOH for Blank}] \times N_{NaOH} \times 0.101 \times 100}{\text{Weight of soil}}$$

[1 ml 1 N H_2SO_4 = 0.101 g of $Ca(HCO_3)_2$]

7.3.7. Estimation of Total Nitrogen in Soil

Total nitrogen in soil samples can be determined by salicylic acid method for the reduction of nitrogenous substances such as nitrite and nitrates into ammonium sulphate, which further converts into ammonia, can be distilled off into boric acid solution and determined by titrimetric method.

a. Reagents

i. Conc. Sulphuric acid

ii. Salicylic acid

iii. Sodium thiosulphate

iv. Potasium sulphate

v. Copper sulphate

vi. **Sodium hydroxide:** 40% solution

vii. **Methyl red indicator:** 0.05% in methanol /water

viii. **Oxalic acid solution (0.1 N):** Dissolve 0.63 g oxalic acid in water and dilute to 100 ml with distilled water.

ix. **Sodium hydroxide solution (0.1 N):** Dissolve 4 g NaOH in water and dilute to 1000 ml with distilled water.

x. **Sulphuric acid solution (0.1 N):** Dilute 3.6 ml conc. H_2SO_4 in water and make up to 1000 ml with distilled water.

b. Procedure

i. Take 4–5 g finely powdered soil sample in a Kjeldahl flask.

ii. Add 30 ml conc. H_2SO_4 containing 1 g salicylic acid, mix thoroughly for 2–3 minutes and allow to stand for about half an hour to complete the reaction where formation of nitrosalicylic acid takes place.

iii. Add 5 g sodium thiosulphate to reduce nitrosalicylic acid into aminosalicylic acid which further produces ammonia. This ammonia is absorbed by sulphuric acid producing ammonium sulphate.

iv. Add 5 g potassium sulphate and 0.5 g copper sulphate into the flask after 5–6 minutes.

v. Digest the content of the flask slowly on a heating device until all the carbon gets oxidized.

vi. After completion of digestion, cool the mixture and dilute to 100–150 ml with water.

vii. Transfer the mixture into a 500 ml round bottom flask.

viii. Pour 40 ml of 40% sodium hydroxide solution into the above mixture and close it immediately.

ix. Distil the content up to one third of its original volume.

x. Collect the distillate (ammonia) evolved during distillation in a measured quantity of standard sulphuric acid/boric acid solution.

xi. Titrate excess sulphuric acid/boric acid solution against standard sodium hydroxide solution using methyl red indicator.

xii. Carry out blank for the same volume.

c. Calculation

Total nitrogen % in soil =

$$\frac{\{\text{Volume of NaOH for blank} - \text{Volume of NaOH for soil sample}\} \times N_{NaOH} \times 0.014 \times 100}{\text{Wt. of soil}}$$

(1ml 1N NaOH = 0.014 g Nitrogen)

7.3.8. Estimation of Ammonium-nitrogen in Soil

Ammonium-nitrogen in soil sample can be determined by converting it into ammonia by the action of alkali solution and thus evolved ammonia is absorbed in dilute sulphuric acid taken in excess. The excess sulphuric acid can be titrated against standard sodium hydroxide solution from which the consumed volume of acid is calculated and ammonium-nitrogen can be determined.

a. Reagents

i. Magnesium oxide

ii. Potassium chloride

iii. **Sulphuric acid (2%):** 2 ml in 98 ml water.

iv. **Standard sulphuric acid solution (0.1 N):** 3.6 ml sulphuric acid diluted to 1000 ml (can be standardised against standard sodium hydroxide solution using phenolphthalein indicator).

v. **Oxalic acid solution (0.1 N):** 0.63 g oxalic acid in 100 ml distilled water.

vi. **Methyl red indicator:** 0.05% in methanol/water.

vii. **Phenolphthalein indicator:** 1% in water/methanol.

b. Procedure

i. Take 10–20 g finely powdered soil sample in a 250 ml Erlenmeyer flask.

ii. Add 5 g magnesium oxide, 0.5 g potassium chloride, 100 ml distilled water into the flask and mix thoroughly.

iii. Start Kjeldahl distillation unit immediately to prevent escape of ammonia.

iv. Shake the flask carefully and distil ammonia using heating device.

v. Collect distillate (ammonia) in 50 ml 2% sulphuric acid solution.

vi. Titrate collected ammonia in excess of sulphuric acid solution against standard sodium hydroxide solution using methyl red indicator.

vii. Cary out blank of same volume of sulphuric acid against same standard sodium hydroxide solution.

viii. Sodium hydroxide can be standardised by titrating against standard oxalic acid solution using phenolphthalein as the indicator.

c. Calculation

Ammonium-nitrogen % in soil =

$$\frac{\{\text{Volume of NaOH for blank} - \text{Volume of NaOH for soil sample}\} \times N_{NaOH} \times 0.014 \times 100}{\text{Wt. of soil}}$$

(1ml 1N NaOH = 0.014 g Nitrogen)

7.3.9. Estimation of Nitrate-nitrogen in Soil

In this method, nitrate is converted to ammonia, which is absorbed in excess dil. sulphuric acid and remaining excess acid can be determined by titrating against standard alkali solution (i.e. sodium hydroxide solution).

a. Reagents

i. **Sodium hydroxide:** (1 %) Dissolve 1 g in 100 ml distilled water.

ii. **Standard sodium hydroxide solution (0.1 N):** Dissolve 4 g NaOH in water and dilute to 1000 ml with distilled water. Standardise it with standard oxalic acid (0.1 N) using phenolphthalein as an indicator.

iii. **Sulphuric acid (2%):** Add 2 ml sulphuric acid in distilled water and dilute to 100 ml.

iv. **Standard sulphuric acid (0.1 N):** 3.6 ml conc. sulphuric acid is diluted to 1000 ml and standardised against standard alkali solution.

v. **Oxalic acid solution (0.1 N):** Dissolve 0.63 g oxalic acid in water and dilute to 100 ml distilled water.

vi. **Devardas alloy** (Cu:50%, AI:45% and Zn:5%)

vii. **Phenolphthalein indicator:** 1% in water/methanol

viii. **Methyl red indicator:** 0.1 % in water/methanol.

b. Standardisation of sodium hydroxide solution against standard oxalic acid solution

   i. Take aliquot of 0.1 N NaOH solution in a 250 ml beaker. Dilute to about 100 ml with distilled water. Add 2–3 drops of phenolphthalein indicator.

   ii. Titrate against 0.1 N oxalic acid solution.

   iii. Disappearance of red colour shows the end point.

Calculation

$$\frac{\text{Normality of oxalic acid} \times \text{Volume of oxalic acid used in titration}}{\text{Volume of NaOH taken (aliquot)}}$$

c. Standardization of sulphuric acid against standard sodium hydroxide acid solution

   i. Take suitable aliquot of sodium hydroxide solution in a 250 ml beaker. Add distilled water to dilute to about 100 ml. Add 2–3 drops of phenolphthalein indicator.

   ii. Titrate against sulphuric acid. Disappearance of red colour shows the end point.

Calculation

Normality of H_2SO_4 $(N_{H_2SO_4})$ =

$$\frac{\text{Normality of NaOH} \times \text{Volume of NaOH used in titration}}{\text{Volume of } H_2SO_4 \text{ taken (aliquot)}}$$

d. Procedure

   i. Take residue remained after the determination of ammonium-nitrogen in the same flask.

   ii. Add 1 g Devardas alloy and 25 ml 1% NaOH into the flask.

   iii. Keep this mixture on a distillation unit and allow to stand for overnight.

   iv. Evolved ammonia is absorbed in 50 ml of 2% H_2SO_4 through condenser of Kjeldahl distillation unit.

   v. Next day, distill the mixture. Allow evolved ammonia to be absorbed in sulphuric acid.

   vi. Take the ammonia absorbed sulphuric acid and 2–3 drops of phenolphthalein indicator.

   vii. Titrate against standard sodium hydroxide solution.

   viii. Disappearance of red colour is the end point.

   ix. Carry out blank titration of the same volume of sulphuric acid taken for the sample.

Calculation

$NO_3 - N$ % =

$$\frac{\{\text{Volume of NaOH for blank} - \text{Volume of NaOH for soil sample}\} \times N_{NaOH} \times 0.014 \times 100}{\text{Wt. of soil sample}}$$

(1ml 1N NaOH = 0.014 g Nitrogen)

7.3.10. Estimation of Available Phosphorus in Soil

A portion of phosphorus, which is utilised by the plant directly, is known as available phosphorus. This can be determined by sodium bicarbonate and acid-fluoride methods.

7.3.10.1. Estimation of Phosphorus in Soil (Olsen-Bicarbonate Method)

a. Reagents

i. **Sodium bicarbonate solution (0.5 M:** Dissolve 42 g $NaHCO_3$ (sodium bicarbonate) in water and dilute to 1000 ml with distilled water, adjust pH = 8.5.

ii. **Sulphuric acid (5 N):** Add 141 ml conc. H_2SO_4 in 500 ml water and dilute to 1000 ml after cooling.

iii. **Stannous chloride solution (0.1%):** Dissolve 0.1 g stannous chloride in 5 ml conc. HCI and dilute to 100 ml with distilled water.

iv. **Ammonium molybdate solution (1% in 4 N H_2SO_4):** Dissolve 10 g ammonium molybdate in 25 ml hot distilled water. Cool and add 110 ml conc. H_2SO_4. Dilute to 1000 ml.

v. **P-nitrophenol solution (0.25%):** Dissolve 0.25 g P-nitrophenol in water and dilute to 100 ml with distilled water.

vi. **Standard stock phosphate solution:** Dissolve 0.4393 g potassium dihydrogen phosphate (KH_2PO_4) in water and dilute to 1000 ml with distilled water. (1 ml = 0.1 mg P).

vii. **Standard phosphate solution:** Dilute 10 ml of stock phosphate solution to 1000 ml with distilled water. (1 ml = 1 mg P).

b. Preparation of standard graph

i. Take suitable aliquots from standard phosphate solution ranging from 0 to 10 mg P in 100 ml volumetric flasks.

ii. Adjust pH = 5 (can be checked using universal pH indicator paper).

iii. Add 5 ml ammonium molybdate solution and 0.25 ml stannous chloride solution in each flask and mix.

iv. Make the volume to 100 ml with distilled water and shake thoroughly.

v. Measure absorbances/transmittances at 660 nm of each sample and blank solution.

vi. Plot graph between absorbances/transmittances against concentrations of phosphorus.

c. Procedure

i. Take 2–5 g of soil sample in a 250 ml conical flask.

ii. Add 100 ml sodium bicarbonate solution into the conical flask. Shake a flask for about 30 minutes.

iii. After 30 minutes filter through Whatman No. 40 filter paper.

iv. If the filtrate is coloured, add a pinch of carbon black. Shake for some time to remove colour and filter again through Whatman No. 40 filter paper.

v. Make up the volume of filtrate 100 ml in a volumetric flask.

vi. Take suitable aliquot from the above prepared solution in a 100 ml volumetric flask. Adjust pH = 5 by adding acid/alkali solutions.

vii. Add 5 ml ammonium molybdate solution and 0.25 ml stannous chloride solution, mix thoroughly.

viii. Make the volume to 100 ml with distilled water and mix thoroughly. Allow to stand for 10 minutes.

ix. Measure the absorbance/transmittance at 660 nm. Carry out blank in a similar manner.

x. Find the concentration of phosphorus from the standard graph.

Calculation

$$P \% = \frac{\mu g \ P \ \text{in aliquot (from graph)} \times 100}{\text{Wt. of sample in gm in aliquot} \times 1000 \times 1000}$$

7.3.10.2. Estimation of phosphorus by Bray's method

This method is suitable for the determination of phosphorus in acidic soil.

a. Reagents

i. **Ammonium fluoride solution (1 N NH$_4$F):** Dissolve 37 g of NH$_4$F in distilled water and dilute to 1000 ml with distilled water.

ii. **Hydrochloric acid (0.5M):** Dilute 40.4 ml conc. hydrochloric acid to 1000 ml with distilled water.

iii. **Extracting solution:** Mix 32.60 ml of 1N NH$_4$F solution and 54.4 ml of 0.5 N HCl solution and dilute to 1000 ml with distilled water.

iv. **Stannous chloride solution:** Dissolve 0.12 g stannous chloride in 2.5 ml conc. hydrochloric acid by warming and dilute to 100 ml with distilled water.

v. **Standard phosphorus solution:** Dissolve 0.4393 g KH$_2$PO$_4$ in distilled water and dilute to 1000 ml with distilled water (1 ml = 0.1 mg P = 100 µg P).

vi. Dilute 10 ml 100 µg P to 100 ml with distilled water. (1 ml = 0.01 mg P = 10 µg P).

vii. **Ammonium molybdate solution (1.5% in 35% HCl):** Dissolve 15 g ammonium molybdate in 350 ml conc. HCl and dilute to 1000 ml with water.

b. Preparation of standard graph

i. Take suitable aliquots from the standard phosphorus solution ranging from 0–10 µg P in 100 ml volumetric flasks.

ii. Add 2 ml ammonium molybdate solution and 1 ml of stannous chloride solution in each flask, mix.

iii. Make up the volume to 100 ml with distilled water, mix thoroughly.

iv. Measure absorbances/transmittances at 660 nm after 10 minutes. Carry out blank in a similar manner.

v. Plot standard graph between absorbances/transmittances and the concentrations of phosphorus.

c. Procedure

i. Take 1–2 g soil sample in a 100 ml beaker. Add 7–14 ml extracting solution, shake for 2 minutes.

ii. Filter through Whatman No. 40 filter paper.

iii. Take the filtrate for determination of phosphorus.

iv. Take suitable aliquot from extracted solution in a 100 ml volumetric flask add 2 ml ammonium molybdate solution and 1 ml stannous chloride solution. Mix well.

v. Make up the volume to 100ml with distilled water, mix thoroughly.

vi. Measure absorbance at 660 nm after 10 minutes. Carry out blank in a similar manner.

d. Calculation

$$P\% = \frac{\mu g\ P\ \text{in aliquot (from graph)} \times 100}{\text{Wt. of sample in gm in aliquot} \times 1000 \times 1000}$$

7.3.11. Estimation of Total Sulphur in Soil

Total sulphur can be determined by converting sulphur into sulphate by oxidizing agent and analysis by colorimetric method.

a. Reagents

i. **Digesting solution:** Dissolve 100 g potassium nitrate (KNO_3) in water and add 350 ml conc. nitric acid and dilute to 1000 ml with distilled water.

ii. **Nitric acid (25%):** 250 ml conc. nitric acid is diluted to 1000 ml with distilled water.

iii. **Hydrochloric acid (6N):** 484 ml conc. hydrochloric acid is diluted to 1000 ml with distilled water.

iv. **Sulphate sulphur (SO_4–S) solution (10 ppm):** Dissolve 0.0544 g potassium sulphate (K_2SO_4) in 1000 ml 6 N HCl.

v. **Barium chloride $BaCl_2$ crystals**

vi. **Standard stock solution:** Dissolve 0.4218 g K_2SO_4 in water and dilute to 1000 ml with distilled water (1 ml = 0.10 mg SO_4–S = 100 μg SO_4–S).

vii. **Standard SO_4–S solution:** Dilute 10 ml 100 μg SO_4–S solution to 100 ml with distilled water (1 ml = 10 μg SO_4–S).

b. Preparation of standard graph

i. Take suitable aliquots of above standard SO_4–S solution in the range of 0 to 100 μg SO_4–S in 100 ml volumetric flasks.

ii. Add 0.25 g barium chloride in each volumetric flask, swirl to dissolve.

iii. Add 1 ml (6N HCL + 10 ppm SO_4–S) solution and allow to stand for 1 minute.

iv. Again swirl the content in the flask.

v. Make up the volume to 100 ml with distilled water.

vi. Measure absorbances/transmittances at 420 nm after 3 minutes.

vii. Plot a graph between absorbances and SO_4–S concentrations.

c. Procedure

i. Take 1–2 gm dried fine soil sample in a 250 ml beaker. Add 10 ml digesting solution.

ii. Keep on water bath and allow to evaporate up to dryness. Replace the beaker and allow to cool.

iii. Add 5 ml of 25% Nitric acid solution. Digest the content for one hour on steam bath.

iv. Dilute to about 50 ml with distilled water.

v. Filter through Whatman No. 40 filter paper.

vi. Transfer the filtrate into 100 ml volumetric flask and make up with distilled water.

vii. Take suitable aliquot of above filtrate or whole solution for determination of SO_4–S.

viii. Add 0.25 gm barium chloride crystal, swirl to dissolve.

ix. Add 1 ml of 6 N HCI + 10 ppm SO_4–S solution. Swirl contents for about I minute.

x. Make up the volume to 100 ml, mix well.

xi. Take absorbance at 420 nm after 3 minutes.

xii. Carry out blank in similar manner.

d. Calculation

$$\% \ SO_4\text{–S} = \frac{SO_4 - S \text{ in aliquot (in µg) from graph}}{\text{Wt. of soil sample in aliquot (in g)} \times 10^7}$$

7.3.12. Estimation of Available Sulphur in Soil

Available sulphur can be determined using turbidimetric/colorimetric method by formation of suspension with barium chloride to barium sulphate.

a. Reagents

i. **Phosphorus extraction solution (500 ppm):** Dissolve 2.195 g potassium dihydrogen orthophosphate (KH_2PO_4) in water and dilute to 1000 ml with distilled water.

ii. **Nitric acid (25%):** 25 ml conc. HNO_3 is diluted to 100 ml water.

iii. **Acetic acid phosphoric acid mixture:** Mixture of 300 ml acetic acid and 100 ml phosphoric acid.

iv. **Gum acacia-acetic acid solution:** Dissolve 5 g gum-acacia in hot water. Cool, filter through Whatman No. 42 filter paper if precipitate appears. Dilute the filtrate to 1000 ml with distilled water.

v. **Standard stock sulphate solution:** Dissolve 0.545 g potassium sulphate in water and dilute to 1000 ml with distilled water. (1 ml = 0.10 mg S = 100 µg S).

vi. **Standard working solution:** Dilute 10 ml 1000 µg S solution to 1000 ml with distilled water. (1 ml = 10 µg S.

vii. **Barium chloride:** $BaCl_2$ crystals.

viii. **Barium sulphate suspension:** Dissolve 20.45 gm of barium chloride in 50 ml hot distilled water. Add 0.5 ml stock sulphate solution, mix. Heat to boiling and cool. Add 5 ml of gum acacia-acetic acid solution, mix again. (This suspension is not stable hence it should be prepared freshly).

b. Preparation of standard graph

i. Take suitable aliquots of standard sulphur solution ranging from 0 to 20 µg S in 25 ml volumetric flasks.

ii. Add 2.5 ml of 25% nitric acid in each flask, mix.

iii. Add 2 ml acetic acid-phosphoric acid mixture, mix well.

iv. Add 0.5 ml barium sulphate suspension, mix well.

v. Add 0.2 gm (or pinch) barium chloride crystals in the mixture and shake for about 5 minutes.

vi. Add 1 ml gum-acacia solution after 5 minutes and make up the volume to 25 ml with distilled water, mix well again. Allow to stand for about ½ hour.

vii. Measure the absorbances/transmittances at 440 nm.

viii. Plot a graph between absorbances/transmittances and concentrations of sulphur.

c. Procedure

i. Take about 20 g dried fine soil sample in a 250 ml conical flask.

ii. Add 100 ml of 500 ppm phosphate extraction solution, shake for about ½ an hour.

iii. Filter through Whatman No.42 filter paper.

iv. Take this filtrate or a part of it for determination of sulphur content in soil.

v. Take suitable aliquot or the total volume (if less than 25 ml) in 25 ml volumetric flask.

vi. Add 2.5 ml of 25% nitric acid solution, mix.

vii. Add 2 ml acetic acid-phosphoric acid mixture, mix again.

viii. Add 0.5 ml barium sulphate acid suspension, mix well.

ix. Add 0.2 gm (pinch of) barium chloride. Shake the contents for 5 minutes.

x. Add 1 ml gum-acacia solution after 5 minutes.

xi. Make up the volume with water and mix well again. Allow to stand for ½ an hour.

xii. Take absorbance/transmittance at 440 nm.

xiii. Carry out blank in a similar manner and deduct from sample reading

Calculation

$$\% \, S = \frac{\text{S in aliquot (in } \mu g) \text{ from graph}}{\text{Wt. of soil sample in aliquot (in g)} \times 10^7}$$

7.3.13. Estimation of Mercury in Soil

Mercury can be determined by using mercury analyser at the trace level in soil samples.

a. Reagents

i. **Aqua-regia:** Mix 1 part of conc. nitric acid and 3 parts of conc. hydrochloric acid.

ii. **Potassium permanganate solution (5%):** Dissolve 5 g $KMnO_4$ (potassium permanganate) in 100 ml distilled water.

iii. **Potassium persulphate solution (5%):** Dissolve 5 gm potassium persulphate $(K_2S_2O_8)$ in 100 ml distilled water.

iv. **Hydroxylamine hydrochloride solution (20%):** Dissolve 20 g hydroxylamine hydrochloride $(NH_2OH–HCI)$ in 100 ml conc. hydrochloric acid.

v. **Stannous chloride solution (20% in 50% HCI):** Dissolve 30 g stannous chloride in 50 ml conc. hydrochloric acid and dilute to 100 ml with distilled water.

vi. **Standard stock mercury solution:** Dissolve 0.1354 g mercuric chloride $(HgCl_2)$ in 2% HNO_3 and dilute to 1000 ml (1 ml = 100 ppm Hg).

vii. Dilute 10 ml 100 ppm stock mercury solution to 1000 ml, add 2% HNO_3 solution (1ml = 1 ppm = 1000 µg Hg).

viii. Dilute 10 ml of 1000 µg Hg solution to 1000 ml by adding 2% HNO_3 solution (1 ml = 10 µg Hg).

ix. OR Dilute 10 ml of 1000 µg Hg solution to 100 ml by adding 2% HNO_3 solution (1 ml = 100 µg Hg).

x. This solution can be taken for standardization.

b. Preparation of standard graph

i. Take suitable aliquot of standard mercury solution ranging from 0–500 µg Hg in a 300 ml B.O.D. bottle (Mercury generator bottle).

ii. Add distilled water so that the total volume of solution should be 100 ml (including all reagents).

iii. Add 5 ml stannous chloride ($SnCl_2$) which is included in 100 ml total volume of BOD bottle.

iv. Place a magnetic bar inside the bottle.

v. Keep the BOD bottle on magnetic stirrer and allow to stir for a fixed period (5 minutes).

vi. Before this check mercury analyser as per the manufacturer's instructions and keep ready for taking the readings.

vii. After a fixed period, i.e. 5 minutes, start the pump and measure the absorbance of mercury shown on the meter.

viii. Feed blank and deduct from every reading.

ix. Plot a graph between absorbances and the concentrations of mercury in µg of Hg

c. Procedure

i. Take about 1–5 gm finely powdered soil sample in a 100 ml beaker. Add 25 ml aqua-regia in a beaker.

ii. Keep for about ½ an hour by swirling occasionally.

iii. Heat on water bath for about 2 hours with swirling occasionally. Cool to room temperature.

iv. Add 25 ml 5% $KMnO_4$ solution, mix.

v. Add 2 ml 5% $K_2S_2O_8$ solution, mix again.

vi. Allow to stand for overnight for reaction.

vii. Next day add 20% hydroxylamine hydrochloride solution drop-wise to oxidise $KMnO_4$ solution completely.

viii. Filter through Whatman filter paper No 40. Wash with distilled water.

ix. Make up the volume to 100 ml or take complete solution for the determination of mercury depending upon the quantity of mercury present in soil sample.

x. Take aliquot or complete solution as prepared above in 300 ml BOD bottle.

xi. Add distilled water to make fixed volume as done in standard graph.

xii. Add 5 ml stannous chloride solution which includes in fixed volume (total volume = 100 ml).

xiii. Place one stirring bar inside the BOD bottle.

xiv. Keep on magnetic stirrer and allow to continue for fixed time (5 minutes).

xv. After five minutes take absorbance on meter.

xvi. Find out the concentration of mercury from the standard graph.

xvii. Carry out blank and deduct from reading.

d. Calculation

$$\% \text{ of Hg in soil} = \frac{\text{Amount of Hg in aliquot (in µg) from graph}}{\text{Wt. of soil sample in aliquot (in g)} \times 10^7}$$

7.3.14. Estimation of Boron in Soil

a. Reagents

i. **Buffer-masking Solution:** Dissolve 250 g of ammonium acetate and 15 g of disodium ethylenediamine tetraacetate, (EDTA disodium salt) in 400 ml of distilled water and slowly add 125 ml of glacial acetic acid.

ii. **Azomethine-H solution:** Dissolve 0.45 g of Azomethine-H in 100 ml of 1 percent L-ascorbic acid solution. Prepare fresh reagent each week and store in the refrigerator.

iii. **Boron stock solution (1,000 ppm boron):** Weigh 5.716 g boric acid (H_3BO_3) into a 1000 ml volumetric flask and dilute to volume with distilled water.

iv. **Boron stock solution (20 ppm boron):** Pipet 20 ml of 1,000 ppm boron solution into 1,000 ml volumetric flask and dilute to volume with distilled water.

b. Preparation of standard graph

i. Pipette 0, 1.0, 2.0, 3.0, 4.0 and 6.0 ml of 20 ppm boron into 100 ml volumetric flasks and dilute to volume with distilled water. Respective amount of boron are 0, 0.2, 0.4, 0.6, 0.8 and 1.2 mg/l. Transfer these to 200 ml beakers.

ii. Add 2 ml of the buffer-masking solution and 2 ml of Azomethine-H reagent to each beaker; thoroughly mix by swirling.

iv. Allow mixture to stand 30 minutes, then measure light transmission at 420 nm wavelength by spectrophotometer.

v. Prepare the standard graph between boron concentrations versus transmittances.

c. Procedure

i. Take about 2 g of dry soil power in a 250 ml beaker, add 2 ml of the buffer-masking solution and 2 ml of Azomethine-H reagent to it; mix thoroughly by swirling.

ii. If necessary, filter supernatant solution through Whatman No. 42 filter paper fitted in plastic funnels.

iii. Inspect filtrate for clarity and refilter if necessary. If filtrate is strongly yellow, refilter with ½ teaspoon of activated charcoal (washed several times with dilute HCl to remove possible boron contamination) in the filter paper cone.

iv. Allow mixture to stand 30 minutes, measure light transmission at 420 nm wavelength by spectrophotometer.

v. Read the concentration of boron in the sample from the standard curve.

7.3.15. Estimation of Chloride in Soil Sample (Mercury Thiocyanate Method)

a. Reagents

    i. **Extracting solution [0.01M Ca(NO$_3$)$_2$·4H$_2$O]:** Weigh 4.72 g into a 2000 ml volumetric flask. Bring to volume with distilled water.

    ii. **Saturated mercury (II) thiocyanate [Hg(SCN)$_2$]solution, 0.075 percent:** Add approximately 0.75 g Hg(SCN)$_2$ to 1000 ml of distilled water and stir overnight. Filter through Whatman No. 42 paper. Since this solution is saturated, it can be stored for long periods of time.

    iii. **Ferric nitrate solution:** Dissolve 20.2 g Fe(NO$_3$)$_3$· 9H$_2$O in approximately 500 ml of distilled water and add concentrated nitric acid (HNO$_3$) until the solution is almost colorless (20 to 30 ml). Make up to 1000 ml with distilled water. Excess HNO$_3$ is unimportant as long as there is enough to prevent darkening of stored solution.

    iv. Charcoal washed in 0.01M Ca(NO$_3$)$_2$ and dried.

    v. **Chloride standard stock solution (1,000 mg/l Cl⁻):** Dissolve 0.2103 g reagent-grade KCl in approximately 50 ml of extracting solution. Bring up to 100 ml.

    vi. **Chloride standard intermediate solution (100 mg/l ppm):** Dilute 10 ml of stock solution to 100 ml with extracting solution.

    vii. **Chloride standard working solutions:** Dilute 0.5, 1.0, 2.0, 3.0, 4.0, 5.0, and 10.0 ml of 100 mg/l standard solution to 100 ml with extracting solution. This is equivalent to 0.5, 1.0, 2.0, 4.0, 5.0, and 10.0 mg/l Cl⁻.

b. Procedure

    i. Take 10 g of powered soil into a 50 ml Erlenmeyer flask. Do duplicate or triplicate analyses. Include a blank. Add approximately 25 mg washed charcoal (dried).

    ii. Add 25 ml extracting solution.

    iii. Shake for 15 minutes at 180 or more excursions per minute (epm) and filter immediately following shaking using Whatman No. 42 filter paper.

    iv. Transfer a 10 ml aliquot to a 50 ml beaker.

    v. Add 4 ml each of the thiocyanate and the ferric nitrate solutions. Swirl to mix.

    vi. Allow 10 minutes for color development and read transmittance or absorbance at 460 nm. Set 100 percent transmittancy with extracting solution.

    vii. Prepare a standard curve by pipetting a 10 ml aliquot of each of the working standards and proceeding as with the soil extracts.

    viii. Plot transmittance or absorbance against concentration of the working standards.

    ix. Determine chloride concentration in the extract from the spectrophotometer reading and standard curve. Subtract the chloride in the blank and convert to ppm in soil by multiplying by a dilution factor of 2.5.

7.3.16. Preparation of Soil for Determination of Trace/Heavy Elements

Finely powdered soil sample is used for the dissolution of heavy elements in aqua-regia solution and can be analysed using appropriate instruments for analysis of them.

a. Reagents

    i. **Hydrochloric acid:** Conc. hydrochloric acid grade.

    ii. **Nitric acid:** Conc. nitric acid.

    iii. **Nitric acid 0.5 M:** Dilute 32 ml conc. HNO_3 to 1000 ml with distilled water.

b. Procedure

    i. Take about 5 g of finely powdered soil sample in a 250 ml beaker.

    ii. Add 21 ml conc. hydrochloric acid, swirl.

    iii. Add 7 ml conc. nitric acid, swirl well again.

    iv. Digest the content on hot plate for about 1–2 hours with covering in fume cupboard.

    v. Cool and filter through Whatman No. 40 filter paper.

    vi. Wash with hot distilled water.

    vii. Collect the filtrate and make up the volume in 100 ml in volumetric flask with distilled water.

Use the above volume for determination of trace/heavy element using appropriate instrument such as Atomic Absorption Spectroscopy, Inductively Coupled Plasma Mass Spectrometry, etc.

Calculation: Calculate the concentrations of each trace/heavy elements and express either in ppm or percentage as applicable.

Analysis of Air Sample

Air is a mixture of different gases out of which nitrogen, oxygen, argon, carbon dioxide and water vapour are most common. In this chapter, analysis of air sample with respect to suspended particulate matter (SPM), SO_2 and NO_2 which degrade the air quality to greater extent has been described. The high volume sampler (HVS) is used to determine the amount of suspended particulate matter (SPM), SO_2 and NO_2 gases present in air.

8.1. SUSPENDED PARTICULATE MATTER (SPM)

The HVS instruments are installed around dust producing sources and measurements are carried out for a fixed period depending upon the concentration of dust so that a sufficient amount of dust is collected on initially weighed filter paper (Whatman glass micro fiber filter paper) GF/A of size 203 × 284 mm. Though the time period can be varied depending upon amount of dust generated, generally 8 hour is considered as optimum period for the collection of suspended particulate matter (SPM).

The volume of air sampled is found out from the initial and final flow rate of air (m^3/min) and time of sampling (in minute).

$$\text{Volume of air sampled in } m^3 = \frac{\text{(Initial flow rate of air) + (Final flow rate of air)}}{2} \times T$$

After collection of sufficient dust, the filter paper is re-weighed and the amount of suspended particulate matter is calculated from the difference in final and initial weights.

$$\text{SPM in } \mu g/m^3 = \frac{\text{(Final wt. of paper + sample) - (Initial wt. of paper)}}{\text{Total volume of air sampled}} \times 10^{-6}$$

8.2. ANALYSIS OF GASEOUS POLLUTANTS

After removal of dust (SPM) the clean air is passed through the impingers containing NaOH and sodium tetrachloromercurate solutions in series so that NO_2 and SO_2 gases are absorbed respectively. The absorbed NO_2 and SO_2 gases in the solutions can be analysed using spectrophotometric method of analysis.

8.2.1. Estimation of Nitrogen Dioxide (NO₂)

Nitrogen dioxide (NO_2) is absorbed quantitatively by bubbling air through 30 ml NaOH solution which forms stable sodium nitrite ($NaNO_2$). By adding appropriate reagents, coloured complex is developed and measurement is taken by spectrophotometer at 540 nm. Concentration of NO_2 in the sampled air is calculated by comparing with standard graph.

a. Reagents

i. **Absorbing reagent:** (0.10 N NaOH): Dissolve 4.0 g sodium hydroxide in distilled water and dilute to 1000 ml .

ii. **Sulphanilamide solution:** Dissolve 20 g sulphanilamide in 700 ml distilled water and add 50 ml of H_3PO_4 (Phosphoric acid 85%) slowly with stirring and dilute to 1000 ml with distilled water. Keep in a refrigerator. This can be used within one month.

iii. **NEDA solution:** Dissolve 1.0 g N (1-naphthyl ethylene diamine dihydrochloride (NEDA) in 1000 ml distilled water. Keep in a refrigerator to protect from sunlight. This solution can be used within one month.

iv. **Hydrogen peroxide:** 1 ml of 30% hydrogen peroxide is diluted to 1000 ml. Keep in a refrigerator to protect from light.

v. **Standard sodium nitrite solution:** Dissolve 0.135 gm sodium nitrite in water and dilute to 1000 ml with distilled water.

b. Standardisation and calibration for NO_2

i. Take 1.0, 2.0, 3.0, 4.0, 5.0, 6.0, 7.0, 8.0, 9.0 and 10.0 ml standard solution of sodium nitrite in 30 ml graduated cylinders.

ii. Add 1ml H_2O_2 solution and 10 ml sulphanilamide solution to each cylinder. Mix thoroughly.

iii. Add 1 ml NEDA solution to each cylinder. Mix thoroughly and make up the volume to 25 ml in each cylinder.

iv. Allow to develop colour for 10 minutes and measure absorbances at 540 nm using spectrophotometer against reagent blank.

v. Plot a graph of concentrations (µg) against absorbance.

c. Procedure for colour development

i. Take 10 ml aliquot from the original solution in a 30 ml graduated cylinder.

ii. Add 1ml H_2O_2 solution and 10 ml sulphanilamide solution. Mix thoroughly.

iii. Add 1 ml NEDA solution. Mix the contents thoroughly, make up volume to 25 ml and allow to develop a colour for about 10 minutes.

iv. Take a blank in the same manner.

v. Measure the absorbance after 10 minutes at 540 nm using spectrophotometer.

vi. Find out the concentration from the standard/calibration graph and calculate the concentration of NO_2/ml.

Calculations

$$NO_2 \text{ in } \mu g/ml = \frac{NO_2 \ \mu g/ml}{10 \text{ ml (amount of aliquot)}}$$

$$NO_2 \text{ in } \mu g/m^3 = \frac{NO_2 \ \mu g/ml \times 30 \text{ (volume of absorbing reagent)}}{\text{Volume of air sampled}}$$

$$\text{Volume of air sampled } (\mathbf{m^3}) = \frac{Q_i + Q_f}{2} \times T$$

Q_i = Initial flow rate of air in m^3/min

Q_f = Final flow rate of air, m^3/min

 T = Total time in minutes

8.2.2. Estimation of Sulphur Dioxide

Sodium tetra-chloromercurate (TCM) absorbs SO_2 quantitatively and forms a dichlorosulphito mercurate complex, after addition of formaldehyde and P-rosaniline at a controlled pH. A coloured complex of P-rosaniline-methylsulphonic acid is formed whose concentration can be measured at 560 nm.

a. Reagents

 i. **Absorbing reagent:** 0.10 M sodium tetra-chloromercurate (TCM): Dissolve 27.2 g mercuric chloride and 11.7 g sodium chloride in water and dilute to 1000 ml with distilled water.

 ii. **Para-rosaniline hydrochloride:** Dissolve 0.2 g of P-rosaniline hydrochloride in water and dilute to 100 ml with distilled water. Allow to stand for 48 hours. Filter if precipitate appears.

 iii. Take 20 ml of P-rosaniline hydrochloride solution, add 6 ml conc. HCI, allow to stand for 5 minutes and make up the volume to 100 ml.

 iv. **Farmaldehyde solution:** (0.2%): Make 5 ml of 40% formaldehyde solution to 1000 ml with distilled water.

 v. **Sulphamic acid solution:** Dissolve 0.6 g sulphamic acid in 100 ml distilled water.

b. Preparation of standard sulphite solution

 i. Dissolve 0.4 gm Na_2SO_3 (sodium sulphite) in water and dilute to 500 ml.

 ii. Take 25 ml aliquot from above solution in a 250 ml conical flask. Dilute to 100 ml with distilled water.

 iii. Add 50 ml of 0.01 N iodine solution and cover immediately and allow to stand for five minutes.

 iv. Titrate the content with 0.01 N sodium thiosulphate solution using starch as indicator.

 v. Change of blue colour to colourless shows the end point of the reaction.

 vi. Carry out a blank in the same manner.

Calculate the concentration of SO_2 in µg/ml using a formula:

$$SO_2 \text{ in } \mu g/ml = \frac{(A - B) \text{ ml} \times N \times 32{,}000}{\text{Aliquot of standard solution}}$$

Where,

 A = Volume in ml of sodium thiosulphate for blank

 B = Volume in ml of sodium thiosulphate for sample

 N = Normality of sodium thiosulphate; 32,000 equivalent for SO_2

c. Preparation of calibration graph

    i. Dilute 2 ml of above standard solution to 100 ml with TMC solution.

    ii. Take aliquots of 1 ml, 2 ml, 3 ml, 4 ml and 5 ml from above solution in graduated cylinders of 30 ml capacity.

    iii. Add 1 ml sulphamic acid solution to each cylinder.

    iv. Add 1 ml P-rosaniline solution to each cylinder.

    v. Add 1 ml formaldehyde solution to each cylinder; mix the contents thoroughly and make up the volume to 20 ml with TMC (tetra-chloromercurate solution). Allow the solution to stand for 10 minutes to develop and slabilise the colour.

    vi. Measure absorbance/concentrations using spectrophotometer at 560 nm against a blank prepared in the same manner.

    vii. Plot a standard graph between absorbance and concentrations.

d. Procedure for colour development

    i. Take 10 ml aliquot from the original solution in a 30 ml graduated cylinder.

    ii. Add 1 ml sulphamic acid solution and 1 ml p-rosaniline solution. Mix well.

    iii. Add 1 ml formaldehyde solution. Mix thoroughly the contents and allow to stand for 10 minutes to stabilise the colour of the complex.

    iv. Measure the absorbance/concentration after 10 min at 560 nm using spectrophotometer. Concentration of SO_2 can be calculated from the standard graph.

Calculations

$$SO_2 \text{ in } \mu g/ml = \frac{SO_2 \ \mu g/ml}{10 \text{ ml (amount of aliquot)}}$$

$$SO_2 \text{ in } \mu g/m^3 = \frac{SO_2 \ \mu g/ml \times 30 \text{ (volume of absorbing reagent)}}{\text{Volume of air sampled}}$$

$$\text{Volume of air sampled (m}^3) = \frac{Q_i + Q_f}{2} \times T$$

Q_i = Initial flow rate of air in m^3/min
Q_f = Final flow rate of air, m^3/min
T = Total time in minutes

Bibliography

1. Bassett, J., Denney, R. C., Jeffery, G. H. and Mendam, J. (1978) Vogel's Textbook of quantitative inorganic analysis including elementary instrumental analysis. ELBS and Longman, New York.
2. Dana, E. S. and Ford, W. E. (1959) A text book of mineralogy. Asia Publishing House, New Delhi
3. Deer, W. A., Howie, R. A. and Zusman, J. (1978) An introduction to the rock forming minerals. ELBS and Longman, London.
4. I.B.M. (2004) Manual of procedure for chemical and instrumental analysis of ore, minerals, ore dressing products and environmental samples. Controller General, IBM, Nagpur.
5. Kelley, W. P., Brown, S. M. and Leibig, G. I. (Jr.) (1940) Chemical effects of saline irrigation water on soils, Soil Science, v.49, pp. 95-107.
6. Read, H. H. (1984) Rutley's Elements of Mineralogy (26th Ed.). CBS Publishers and Distributors, New Delhi.
7. Sawyer, C. N. and Mc Carty, P. L. (1967) Chemistry for sanitary engineers, 2nd Ed. Mc. Grow-Hill, New York, 518p.
8. Sen, A. K. (1995) Laboratory manual of Geology. Modern Book Agency Pvt., Ltd, Kolkata.
9. Shapiro, L. and Brannock, W. W. (1962). Rapid analysis of silicate, carbonate and phosphate rocks. U. S. Geological Survey Bulletin, No. 1144-A.
10. Sharma, Y. R. and Das, A. K. (2007) Practical chemistry. Kalyani Publishers, Cuttack.
11. Subba Rao, D. V. (2003) Coal-its beneficiation. Em Kay Publications.
12. Systronics- Instruction manual of flamephotometer – 104.
13. Systronics- Instruction manual of ion/pH meter – 363.
14. Systronics- Instruction manual of spectrophotometer – 104.
15. Systronics- Instruction manual of water analyzer – 371.

Index